Southern England's
1,000-Foot Peaks

Jeff Kent

WITAN BOOKS

Jeff Kent was born in 1951 and educated at Hanley High School in Stoke-on-Trent. He gained a second-class honours degree in International Relations from the University of London in 1973 and lectured at several colleges in North Staffordshire.

He is a prolific writer on a variety of subjects and has previously published fifteen books: *The Rise And Fall Of Rock*; *Principles Of Open Learning*; *The Last Poet: The Story Of Eric Burdon*; *Back To Where We Once Belonged!: Port Vale Promotion Chronicle 1988-1989*; *The Valiants' Years: The Story Of Port Vale*; *Port Vale Tales*; *Port Vale Forever*; *The Port Vale Record 1879-1993*; *Port Vale Personalities: A Biographical Dictionary of Players, Officials and Supporters*; *The Potteries Derbies*; *The Mysterious Double Sunset*; *What If There Had Been No Port In The Vale?: Startling Port Vale Stories!*; *Staffordshire's 1,000-Foot Peaks*; *Peak Pictures* and *Cheshire's 1,000-Foot Peaks*.

Jeff was the editor of and the main contributor to *The Mercia Manifesto: A Blueprint For The Future Inspired By The Past*; *A Draft Constitution For Mercia* and *The Constitution Of Mercia*. He also edited Denis Dawson's autobiography, *Port Vale Grass Roots: from Supporter to Groundsman and Back Again*; Cyril Kent's autobiographies, *A Potteries Past* and *Stories From Stoke*, and Max F. Sabian's *Dynamic and Instinctive Thinking*. In addition, Jeff was the historical adviser to the Port Vale video histories, *Up The Vale!: The Story of Port Vale FC* and *The Millennium Documentary*, and was the director, screenplay writer, narrator and soundtrack composer of the film *Pictures From The Potteries*.

He was a pioneer of Green music, has performed benefit concerts for several environmental and humanitarian organisations and has released four albums: *Tales from the Land of the Afterglow, Parts 1 & 2*; *Port Vale Forever* and *Only One World*. Consequently, his biography has gained entry into the *International Who's Who In Music Volume Two: Popular Music*. He played percussion in the Glorishears of Brummagem morris dance band from 2008 to 2013 and in the Mercia Morris dance band from 2013 to 2015.

Jeff was the convener of the Confederation for Regional England from 1999 to 2000 and of the Mercian Constitutional Convention from 2001 to 2003. Also, he has been the co-ordinator of The Mercia Movement since its formation in 1993 and the convener of The Acting Witan of Mercia since its foundation in 2003.

Witan Creations
2018 Main Catalogue

WTN 001: *Butcher's Tale/Annie, With The Dancing Eyes* – Jeff Kent & The Witan (animal rights protest single), 1981 - £2.50, including p & p.

WTN 027: *The Mercia Manifesto: A Blueprint For The Future Inspired By The Past* – The Mercia Movement (128-page radical political manifesto), 1997 - £6.00, including p & p.

WTN 030: *Only One World* – Jeff Kent (13-track environmental concept CD), 2000 - £11.25, including p & p.

WTN 032: *A Draft Constitution For Mercia* – The Mercia Movement (20-page draft constitution for an independent Midlands), 2001 - £2.50, including p & p.

WTN 038: *The Constitution Of Mercia* – The Mercian Constitutional Convention (22-page constitution of independent Mercia), 2003 - £2.50, including p & p.

WTN 083: *Pictures From The Potteries* – Cyril and Jeff Kent (115-minute DVD showing the highlights of 131 cine films shot mainly in and around Stoke-on-Trent from 1962 to 1988), 2014 - £11.90, including p & p.

WTN 084: *Peak Pictures* – Jeff Kent (137-page colour photo record of unusual scenes shot in the southern Pennines between 2012 and 2014), 2014 - £7.25, including p & p.

(All other publications and releases by Witan Creations are now out of print or unavailable from the catalogue.)

Southern England's 1,000-Foot Peaks

First published in March 2018 by Witan Books, Cherry Tree House, 8 Nelson Crescent, Cotes Heath, via Stafford, ST21 6ST, England.
Telephone: (01782) 791673.
E-mail: witan@mail.com
Website: www.witancreations.com

Copyright © 2018, Jeff Kent.

WTN 109

ISBN 978 0 9927505 3 4

The right of Jeff Kent to be identified as the author of this work has been asserted by him in accordance with the Copyright, Designs and Patents Act 1988.

A Cataloguing in Publication record for this book is available from the British Library.

Design and cover concept: Jeff Kent.
Cover artwork and digital design: Steve Billington.
Editorial and technical adviser: Sue Bell.
Main research: Jeff Kent.
Printing co-ordinator: Allan Staples.
Printed and bound by: Hanley Print Services, Unit 78, Shelton Enterprise Centre, Bedford Street, Shelton, Stoke-on-Trent, ST1 4PZ. Tel. (01782) 280028/286358; E-mail: info@hanley-print.co.uk.

All rights reserved. No part of this publication may be reproduced, stored in a retrieval system or transmitted in any form or by any means, electronic, mechanical, photocopying, recording or otherwise without the prior permission in writing of the publisher and the copyright holder.

Dedicated to the Southern England peaks baggers.

ACKNOWLEDGEMENTS

I should like to express my thanks to the following people and organisations for their invaluable assistance, not credited elsewhere, in the production of this book: David Edwards, Lesley Flood, Mark Greaves, Ben Humphries, Ordnance Survey, David Pratt and Mark Ratkai.

PHOTOGRAPHS

Front cover photo: Brent Tor, in Devon.
Back cover photo: The Cheesewring, on Stowe's Hill, in Cornwall.
Cover background photo: Rough Tor, in Cornwall.

PHOTOGRAPH CREDITS

Photographs were kindly supplied by the following:
Ian Wool - front cover; Joshua Wells - back cover; Jeff Kent - cover background.

Contents

Preface. 1
1 The Determination Of The Peaks. .3
2 Southern England's Peaks 5
3 Cornwall's 1,000-Foot Peaks 7
4 Devon's 1,000-Foot Peaks 17
5 Somerset's 1,000-Foot Peaks 80

Preface

This book is hopefully the first in a planned series of volumes of all the 1,000-foot peaks of England and its origin lies in the publication of a list of Scotland's 3,000-foot mountains by Hugh Munro in 1891. Although I wasn't aware of that when I visited the Scottish Highlands for the first time on a touring holiday with my parents in 1969, I fell in love with the wild landscape which we saw and returned to Scotland many times.

As it transpired, my favourite part of the country was western Sutherland, where I discovered a number of fantastically-shaped hills, especially the domed and horned Suilven, standing well below Munro level. As a result of my interest in them, I became aware that they were known as Corbetts (peaks with a height between 2,500 and 2,999 feet) and Grahams (those between 2,000 and 2,499 feet). As my knowledge expanded, I encountered the Nuttalls and Hewitts, separate and different lists of the mountains of England and Wales of 2,000 feet and above.

There are no peaks that high in Staffordshire, where I live, but a fair amount of land in the north of the county reaches 1,000 feet and I became pretty familiar with much it from January 1963 when I survived a reckless school hike around The Roaches in bitterly cold weather. Over time, I built up a series of beautiful walks in the area, especially in and around the Manifold Valley, but by the beginning of this century sadly I'd become bored with them. I needed a new landscape challenge and started wondering how many 1,000-foot peaks (that is those between 1,000 feet and 1,999.99 feet in height) there were in the shire. I thought it very unlikely that anyone had ever compiled a list of them and my research proved me to be right. So a new project was born and I decided to produce a list of them myself and to climb all the peaks I discovered. This I did, in 2012, identifying a surprising 65 such hills from Ordnance Survey maps and then conquering them before repeating the exercise with Cheshire's 46 1,000-foot peaks in 2014. I followed up by writing and publishing a book on both, in 2013 and 2015 respectively, and called the hills of each county collectively the *Kents*, in honour of my parents, Cyril and Helen, who first took me into the countryside when I was a young boy and kindled my love of the great outdoors.

Having got the 1,000-foot peaks' bug, there was no stopping me, so in 2015 and 2016, I identified 197 hills of the magic four-figure height in Shropshire and climbed all of them. Last year, I switched my attention to Herefordshire's 44, Worcestershire's eight and Gloucestershire's five 1,000-foot hills and bagged them too, which increased my total number of ascents of such separate peaks in England to 366 in six years. In due course, I aim to produce a book on the Shropshire four-figure hills and another on those of the other three counties combined.

I was accompanied on almost all my ascents by my partner, walking companion and faithful photographer, Sue Bell. She herself reached the vast majority of the precise hill tops and all the summit areas except for that of Harley's Mountain, in Herefordshire, when she became unwell partway through a figure of eight three-peak circular walk on 18 May 2017. After ascertaining that she was simply off colour, I sadly climbed the final hill of the day alone.

In the autumn of 2015, as an extension of my individual counties projects, I decided to take on the huge task of identifying all the 1,000-foot hills in England and then hopefully publish my findings in a single volume. Over the next two years,

I discovered huge numbers of peaks and at times the endeavour seemed endless! The numbers of pages of text grew until they approached 800 and I eventually realised that my book would need to be divided into six volumes of manageable size, split into five different regions of the country, with a concluding and unifying tabular tome of the *Kents* overall. By analysing the breakdown of my text and considering viable and recognisable English regions, I was able to split the book into new sections relatively easily and so *Southern England's 1,000-Foot Peaks* was born, covering the four-figure hills in that part of the country.

As most of my groundwork has now been done, it is my intention that its sister volumes should follow fairly quickly, based on the same formula, so that the whole project can be published within the next couple of years. But the future is unpredictable and time will tell if my aim and its outcome coincide.

From the outset of writing this book and its forthcoming companions, I decided that the mass of peaks awaiting identification needed to be subdivided into sections to make the great volume of text more manageable. I considered the option of presenting them in hill ranges, but that was problematic because the boundaries of such are arguable and some peaks are isolated or relatively so. Therefore it seemed best to present the hills within counties, the borders of which are precisely known. Because of the complexity of the current local government administrative areas in England and their likely changeability, I concluded that the ceremonial, or geographic, counties, as defined in the 1997 Lieutenancies Act, would be the best units to use.

Ideally, I would have included maps in the book showing the location of the peaks, but, unfortunately, I realised that the sheer numbers of them concentrated in relatively small areas, especially on Dartmoor, would create such a clutter as to make the maps practically useless. So, reluctantly, I desisted, but it should still be relatively easy for readers to pinpoint the position of the peaks on appropriate maps from the six-figure grid references I've listed for the summit of every hill. Conversely, as I've arranged the 1,000-foot peaks in descending altitude, those found on such maps can quickly be detected in the book from their shire and metric height, even though the latter might be an approximate match. As the OS 1:25000 map was the main source of information for the book, it is consequently the best one to use in conjunction with it.

Essentially, I've used the names of the peaks as they are shown on the 1:25000 map, but where such are absent, I tried to discover them through other sources. However, some remained unnamed, so I took the liberty of assigning the most appropriate names to them, usually derived from significant nearby features.

Finally, I have parenthesised peaks without a relative height of fifty feet in square brackets and used round ones for those which were previously unnamed or have alternative names.

In completing this book, I did everything I could to ensure that its contents were all accurate at the time of writing. Having said that, errors can be made and things invariably alter over time so that this volume cannot be a bible, although I hope it will become and remain a valuable reference work.

Jeff Kent,
Cotes Heath,
March 2018.

1 The Determination Of The Peaks

In England, there are 256 Nuttalls, compiled in a list by John and Anne Nuttall, which are hills of 2,000 feet and above, with a relative height (or prominence) of at least 15 metres. However, the Nuttalls are rivalled by the Hewitts, produced by Alan Dawson, the main difference being that the latter record uses a prominence of 30 metres or more to determine an independent peak.

I took the above lists, particularly that of the Nuttalls, as my models for the determination of England's (and therewithin Southern England's) 1,000-foot hills, not least because there has to be some form of criteria to decide what an independent peak constitutes. Nevertheless, I selected the precise criteria for inclusion in my record of the country's 1,000-foot peaks myself, according to all the factors which came to my attention:

Firstly, I decided that all the summits included must be natural features (or those topped by tumuli) within the boundaries of England in this series of books, and those of Southern England for this particular volume, and be between 1,000 and 1,999.99 feet above mean sea level.

Secondly, I thought that the best formula to apply in terms of determining an independent hill was an approximation of the one used by the Nuttalls, that is those with a relative height of 15 metres/49.212 feet. (The relative height of a hill is the vertical distance between its summit and the lowest contour line encircling it, within which there is no higher peak.) However, as I was producing a list of 1,000-foot peaks, with a record primarily based on imperial measurements, I decided that the appropriate qualifying relative height to use was 50 feet or 15.24 metres.

As the cutoff point for John and Anne Nuttall was 15 metres for peaks of 2,000 feet in height, then it seemed reasonable that it should not be more for those of 1,000 feet. If I'd chosen a smaller prominence, it would have meant that quite a number of relatively insignificant features would have been included, so that the impression of a peak would have been considerably diluted. Alternatively, I could have gone for a higher relative height, but by doing so I would have reduced the numbers of qualifying hills, by increasing quantities as the prominence level rose. Indeed, had I adopted the Hewitts 30 metres criterion, the current list would have been cropped heavily and some quite significant hills would have been excluded. So a relative height of 50 feet seemed to be the most reasonable amount of prominence required to ascertain 1,000-foot peaks of importance.

Thirdly, I decided, with some reservation, that England's (and thereby Southern England's) 1,000-foot hills named on relevant Ordnance Survey maps which have a summit, but do not satisfy the relative height criterion, would also be included in the list, for example Rough Tor, in Devon, which is actually an outlier of Cut Hill, and Smallacoombe Downs, in Cornwall, whose parent peak is Newel Tor. My reasoning behind this decision was that features which have sufficient total height to qualify and are already named are generally known and therefore their exclusion might appear odd. However, to ensure the existence of a summit on these hills, I decided that they needed to have a minimum relative height of five feet, although even then occasional anomalies resulted!

Excluded from the list are places named on maps as hills, but which contain no summits, and man-made features (except for tumuli), such as buildings, tips, landscaped tips, reservoirs and quarries. Also, the word "Moor" refers to a tract of unenclosed ground and/or an area of heathland, so such names are not regarded as indicating a hill. Where the word appears in a name on a qualifying unnamed peak, my rule of thumb has been simply to change the word "Moor" to that of "Hill" for appropriateness.

Included in the list are features which contain the words "Bank", "Rock(s)" or "Head" in their names when clearly referring to hills or their summits, but they are excluded when indicating slopes without tops, low-standing rock fields on the flanks of hills and the sources of rivers and valleys, respectively. In addition, the word "Low" can mean a mound or a hill. When used in a name, it's taken as referring to a prehistoric feature where one is present, but otherwise it's regarded as indicating a hill and is therefore automatically included in the tally.

To compile the list of England's 1,000-foot peaks, and therefore that of Southern England's, I painstakingly trawled and scrutinised all the innumerable grid squares of the Ordnance Survey's 1:25000 online map which had the potential to contain qualifying hills. I was looking for summits with an altitude of between 304.8 metres (1,000 feet) and a fraction below 609.6 metres (2,000 feet), but initially gave a little leeway either side because of possible discrepancies. As the height of many peaks isn't recorded on OS maps, whenever I found such features, I used the OS Maps website altitude facility to measure their heights, both in feet and metres, and included them if they met the criteria. I recorded the grid references of their summits and then added further information about them from a variety of reliable sources, including my own personal knowledge.

In such a huge undertaking, the chances of accidentally missing out an odd peak or two were considerable, so I trawled the whole area a second time on the OS map and checked numerous other sources, especially a series of hill lists available in printed books and online, to try to find any omissions, of which I fortunately discovered only a few. Realistically, I don't think I could have been any more thorough and I can only hope that no qualifying peaks have escaped my attention.

Unfortunately, for reasons which the OS hasn't been able to explain to me, the heights and summit locations of the 1,000-foot peaks on the OS 1:25000 map and those computed by the OS Maps website often show minor discrepancies and occasionally significant ones. As a result, I have recorded the vast majority of these variations and all the notable ones in the text, with likely explanations where necessary. It is possible to resolve these discrepancies through the ascertainment of precise and virtually definitive altitudes and positions of the hill tops, but this can only be achieved through carrying out painstaking, on-the-spot, professional surveys, which is a task beyond my capabilities and time and therefore is for others to undertake.

I'm well aware that the record of Southern England's 1,000-foot peaks as determined by me is open to criticism as hill x is excluded or ridge y included, or indeed because my selection criteria are too tough or weak, but it is a list nevertheless and the only one that currently exists. However, its real value will be only revealed by the test of time. Perhaps it will remain intact and become definitive, maybe it will be amended and possibly it will be made obsolete by metrication or scrapped in favour of something new which is considered superior, but anyway round it has drawn a visible line across the earth.

2 Southern England's Peaks

There is no agreement as to what constitutes Southern England, even though the notion of it as a geographical region is clear, most especially regarding the perceived South versus North divide between the country's rich and poor. For nearly 250 years prior to the Norman Conquest, the Anglo-Saxon shires south of the Rivers Thames and Avon and the Severn Estuary comprised the Kingdom of Wessex and today the same region (or thereabouts) is frequently divided into the Southeast and Southwest, giving the common denominator that together they make up the South. England's longest river, the west to east-flowing Thames, in particular, has long represented a border, not least in the mind, between the gentile South and the more earthy Midlands to the north of it.

I decided to adopt the same view of the location of Southern England, partly because of the aforementioned geographical and historic reasons, but more significantly as it thus makes sense as a region in the context of England's 1,000-foot peaks and their presentation in book form. North of the Thames, or broadly so, is Central England, most commonly known as the Midlands and historically as Mercia, and this region contains well sufficient 1,000-foot peaks to be considered as a separate entity from Southern England.

The book could just as easily have been called *Southwestern England's 1,000-Foot Peaks*, as no hills outside the counties of Cornwall, Devon and Somerset reach the magic four-figure height, but the consequence of that would have been to have eliminated the Southeast from inclusion within the country's uplands, with the impression given that it lacks any high ground, which is definitely not true.

Of the seven remaining shires of Southern England, the highest peak in no fewer than six of them (Berkshire, Wiltshire, Surrey, Hampshire, Sussex and Dorset) stands at over 900 feet and in the final one (Kent), it surpasses 800 feet. So, although there are only 1,000-foot peaks in three counties in the South, the height of the ground comes close to achieving that figure in the rest of them.

The highest point in Berkshire is the summit of Walbury Hill, on the Hampshire Downs, which is marked on the OS 1:25000 map as reaching 297 metres at the trig point at grid reference SU 373616 and is calculated by OS Maps as rising to 973.75 feet just to the southwest, at SU 372615. Nearby Combe Hill and Inkpen Hill aren't much lower and a further significant area of high ground, the Berkshire Downs, lies to the north.

According to OS Maps, Wiltshire's highest peak, Milk Hill, at the southern end of the Marlborough Downs, attains a height of 963.25 feet at grid reference SU 104643, although the OS 1:25000 map's 295-metre contour ring and a 2009 independent survey (calculating 294.26 metres) indicate it to be slightly higher. Also, its sister peak, Tan Hill, a little to the west, is computed to be less than a foot lower.

Surrey may conjure up images of leafy suburbia, but its highest peak, Leith Hill, on the Greensand Ridge, is given a spot height of 294 metres on the OS 1:50000 map, at the base of its crowning eighteenth century tower at grid reference TQ 139431, and is computed by OS Maps to be 962.60 feet high. Although the impressive South Downs chalk hills are the dominant upland feature of Sussex, this county's summit is also provided by the Greensand Ridge, on Black Down, which is shown on the OS 1:25000 map as standing at 280 metres at its trig point, at grid reference SU 919295, and is reckoned by OS Maps to have an altitude of 919.29 feet.

Approximately 1¾ miles east-southeast of Walbury Hill is situated Hampshire's highest peak, Pilot Hill, which is marked on the OS 1:25000 map as reaching 286 metres at its trig point, at grid reference SU 398600, and is calculated by OS Maps to be 938.65 feet high. Like its near-neighbour, the peak is located on the Hampshire Downs, although its summit is a mere 175 yards from the county boundary with Berkshire.

Dorset reaches its apex at Lewesdon Hill, on the North Dorset Downs, at grid reference ST 437012, at or near a somewhat unclear OS 1:25000 map 279 metre spot height. OS Maps, however, computes its summit to be 2.80 metres lower, at 906.17 feet high. A second peak, Pilsdon Pen, approaching 1½ miles to the west, attains an only slightly lower altitude, of 277 metres, according to the OS map and 903.22 feet, as determined by OS Maps.

Finally, the highest peak in Kent, Betsom's Hill, is located on the North Downs. Its summit is shown on the OS 1:25000 map as being within a 245-metre contour ring and is reckoned by OS Maps to stand at 249.80 metres, which is 819.55 feet. Although it's clearly not close to being a 1,000-foot peak, it's a sizable hill nevertheless.

The most easterly 1,000-foot peaks in Southern England are situated on the gentle limestone Mendip Hills, in northern Somerset, but the vast majority of the 408 four-figure peaks of the region are found in the Southwest Peninsula, to the west of a line drawn from Lyme Bay to Bridgwater Bay. As the peninsula is mainly fairly low lying, the bulk of these peaks are concentrated in several ranges of high hills, especially Dartmoor (in Devon), Exmoor (which is split between Somerset and Devon), Bodmin Moor (in Cornwall), the Brendon Hills (in Somerset) and the Quantock Hills (also in Somerset). In turn, the hilly areas are split into two distinct types: the generally smooth-profiled sedimentary ones of Somerset and northern and eastern Devon and the granite moorlands, with outcropping bare rock tors, in southern and western Devon and Cornwall. The latter, and particularly their frequent tor summits, tend to be more dramatic and rugged.

Overall, Southern England consists of low ground punctuated by intruding hill ranges, mostly gently in nature, but higher and more impressive in the west of the region (especially in Devon). It is there that Southern England's 1,000-foot peaks are located.

3 Cornwall's 1,000-Foot Peaks

Although there are several upland areas in interior Cornwall, the largest and easily the highest is the granite Bodmin Moor, towards the eastern end of the county, where the vast majority of the 1,000-foot peaks are found. A few scattered hills beyond the moor reach a four-figure height, but nearly all of them are at no great distance from it. No point in Cornwall rises to anywhere near 2,000 feet and Brown Willy, on Bodmin Moor, is the county's highest hill, as well as being its tallest 1,000-foot peak.

1. Brown Willy, 1,361.55 feet/415.00 metres, SX 158799, an elongated, partly rocky peak, which is the highest hill in Cornwall. It rises steadily and then quite steeply from the De Lank Valley to the west, very gently and then steeply from the Fowey Valley to the east and variably from a lake (via Brownwilly Downs) to the south and Roughtor Marsh to the north-northwest. OS Maps shows its rocky summit to be by the OS 1:25000 map trig point, which is marked as standing at 420 metres. The sizable discrepancy in the peak's altitude given by the online and paper versions of the OS's data might be explained by a limitation of the website's programme to calculate accurately the heights of rapidly-rising features. There are quite a number of prehistoric sites on its slopes. High Moor is situated on its north-northeastern flank and Maiden Tor on its northern one.

2. Rough Tor, 1,286.09 feet/392.00 metres, SX 145808, a somewhat elongated, very rocky peak, which rises increasingly steeply from a tributary valley of the River Camel (via the extensive Stannon [china clay] Works) to the west, very gently and then fairly steeply from a col (via Roughtor Moors) to the south and variably from the De Lank Valley to the east and Crowdy Reservoir to the north. OS Maps shows its rocky summit to be at the OS 1:25000 map 400-metre spot height. The large discrepancy in the peak's altitude given by the online and paper versions of the OS's data might be explained by a limitation of the website's programme to calculate accurately the heights of rapidly-rising features. Numerous prehistoric and medieval remains lie on its slopes, including Fernacre Stone Circle to the south. Little Rough Tor and Showery Tor are situated on its upper north-northeastern flank and an area of woodland arcs around its northern one.

3. Kilmar Tor (High Rock), 1,273.62 feet/388.20 metres, SX 251748, a 1,000-yard-long, very rocky, undulating, west-southwest to east-northeast-running ridge. It rises increasingly steeply from the Withey Valley to the west and cols to the north-northwest and southeast and variably from the Lynher Valley to the east. OS Maps shows its summit to be on the top of High Rock, approximately 200 yards west-southwest of a trig point, although the OS 1:25000 map and other sources indicate that a rock tor about 50 yards southwest of the pillar (marked on the map as 396 metres) is higher. The large discrepancy in the peak's altitude given by the online and paper versions of the OS's data might be explained by a limitation of the website's programme to calculate accurately the heights of rapidly-rising features. There are numerous prehistoric sites on its slopes and most of its upper ones are covered by Twelve Men's Moor.

4. [Little Rough Tor], 1,258.20 feet/383.50 metres, SX 148810, a somewhat elongated, very rocky peak on the east-northeastern part of the Rough Tor ridge. It rises fairly steadily and then quite steeply from the De Lank Valley to the east, fairly steadily from a tributary stream of the River Camel to the west and fairly gently from a col to the north-northeast. OS Maps shows its summit to be at the OS 1:25000 map 390-metre spot height. The considerable discrepancy in the peak's altitude given by the online and paper versions of the OS's data might be explained by a limitation of the website's programme to calculate accurately the heights of rapidly-rising features. The hill doesn't have a relative height of fifty feet and is linked to its parent peak by a very high col. There are several prehistoric sites on its slopes.

5. [Showery Tor], 1,246.06 feet/379.80 metres, SX 149813, a roundish, pretty rocky peak on a north-projecting spur of Rough Tor (and, more immediately, Little Rough Tor), to the south-southwest. It rises gently and then fairly steadily from the De Lank Valley and Roughtor Marsh to the east, fairly gently from a tributary stream of the River Camel to the west and variably from Crowdy Reservoir to the north. OS Maps shows its summit to be on the top of an impressive outcrop of rocks within the OS 1:25000 map large Bronze Age cairn, which is marked as reaching 385 metres. The fair discrepancy in the peak's altitude given by the online and paper versions of the OS's data might be explained by a limitation of the website's programme to calculate accurately the heights of rapidly-rising features. The hill doesn't have a relative height of fifty feet and is linked to its parent peak by a high col. There are several prehistoric sites on its slopes; a medieval long-house is situated on its east-southeastern flank and an area of woodland curves around its northern one. The source of the De Lank River is located at the foot of its northeastern slopes, within Roughtor Marsh.

6. Langstone Downs, 1,240.49 feet/378.10 metres, SX 255737, an ovalish-shaped, partly rocky peak, which rises fairly steadily and then very gently from a col to the east, quite gently from another to the south, quite gently and then gently from the Withey Valley to the west and very gently from another col to the north. OS Maps shows its summit to be just east-southeast of the OS 1:25000 map 379-metre spot height, although other evidence indicates it to be on the edge of the most easterly of three prehistoric cairns. There are numerous prehistoric sites on its slopes; Sharp Tor and the hamlet of Henwood are situated on its east-southeastern flank; Boundary Rock is located on its east-northeastern one and Bearah Tor is sited on its north-northeastern one.

7. Stowe's Hill, 1,237.20 feet/377.10 metres, SX 257724, an elongated, pretty rocky peak, which rises increasingly steeply from the head of the Withey Valley to the west and a col to the north-northwest and variably from a tributary dale of the River Lynher to the east and the head of the Seaton Valley to the south-southeast. OS Maps shows its rocky summit to be at the south-southwestern end of the OS 1:25000 map 380-metre tiny contour ring. The Stowe's Pound Neolithic enclosures encircle most of its greater summit area, on which are also situated two prehistoric cairns and the impressive Cheesewring tor. The disused Cheesewring

Quarry is located within its southern enclosure and several disused tips and shafts and one or two disused mines and quarries are sited on its slopes. The (three) Hurlers stone circles and The (two) Pipers standing stones can be seen on its southern flank and the village of Minions and the source of the River Seaton are found on its south-southeastern one.

8. Caradon Hill, 1,211.94 feet/369.40 metres, SX 272707, a wedge-shaped, partly rocky peak, which rises quite gently from the Seaton Valley to the west, variably from one of its tributary dales to the south and one of the River Lynher to the east and fairly steadily from another of the latter to the north. OS Maps shows its summit to be at the OS 1:25000 map trig point, which is marked as standing at 369 metres, but the map and other sources indicate that its highest spot is on the top of the highest of several prehistoric cairns situated on its summit area and give it a height of 371 metres. A television station and three telecommunications masts are located on its summit area; there are numerous prehistoric cairns and disused tips and several disused mines and shafts on its slopes; the source of the River Seaton is sited at the foot of its west-northwestern flank and that of the River Tiddy can be found at the bottom of its south-southeastern one.

9=. The Beacon (Hendra Downs), 1,206.04 feet/367.60 metres, SX 196792, a somewhat elongated, slightly rocky, roundish-topped peak, which rises fairly steadily from a tributary dale of the River Lynher to the east, gently and then quite gently from the Fowey Valley to the west and variably from the latter to the south and a tributary valley of Penpont Water to the north. OS Maps shows its summit to be on the top of the more easterly of two OS 1:25000 map prehistoric cairns to the east of the 369-metre spot height. The Elephant Rock is situated on its south-southwestern flank, Black Rock on its northern one and Webb's Down on its south-southeastern one. There are several prehistoric sites on its slopes and the A 30 runs across its lower southern and eastern ones, broadly from southwest to northeast.

9=. [Sharp Tor], 1,206.04 feet/367.60 metres, SX 260736, a small, very rocky peak on an east-southeast-projecting spur of Langstone Downs. It rises steadily from a col to the east, fairly steadily from a dry valley to the south and fairly gently from another to the north. OS Maps shows its rocky summit to be just south-southwest of the OS 1:25000 map 378-metre spot height. The large discrepancy in the peak's altitude given by the online and paper versions of the OS's data might be explained by a limitation of the website's programme to calculate accurately the heights of rapidly-rising features. The hill doesn't have a relative height of fifty feet and is linked to its parent peak by a high col. The village of Henwood is situated on its east-southeastern flank and a prehistoric field system on its north-northeastern one.

11. [Bearah Tor], 1,185.70 feet/361.40 metres, SX 257744, a small, very rocky peak on a north-northeast to east-northeast-curving spur of Langstone Downs to the south-southwest. It rises quite gently from a depression to the south, very gently and then steadily from another to the north and a col to the west-northwest and variably from the Shales Valley to the east. OS Maps shows its rocky summit to be just south-southwest of the OS 1:25000 map 367-metre spot height. The fair discrepancy in the peak's altitude given by the online and paper versions of the OS's data might be explained by a limitation of the website's programme to calculate accurately the heights of rapidly-rising features. The hill doesn't have a relative height of fifty feet and is linked to its parent peak by a high col. Bearah Tor Quarry is situated on its upper eastern flank and there are several prehistoric remains on its slopes.

12. Tolborough Tor (Tolborough Downs), 1,137.47 feet/346.70 metres, SX 175778, a roundish, slightly rocky peak, which rises increasingly steeply from a tributary valley of the River Fowey to the east and gently from another to the south, dry dales to the north and west and a col to the west-northwest. OS Maps shows its highest point to be on the east-northeastern edge of the OS 1:25000 map summit prehistoric cairn, which is marked as having a height of 348 metres.

13. Bray Down, 1,132.87 feet/345.30 metres, SX 189821, a wedge-shaped, partly rocky, roundish-topped peak, which rises fairly steeply and then quite gently from a tributary dale of the Penpont Valley to the east, variably from the head of the same stream to the south, fairly steadily from Penpont Water to the west and gently and then steadily from the latter to the north. OS Maps shows its highest point to be by the OS 1:25000 map trig point, which is marked as standing at 346 metres. Two prehistoric cairns are situated on its summit area; Bray Down Iron Age enclosure is located on its north-northeastern flank and a prehistoric settlement is sited on its west-southwestern one.

14. Catshole Downs, 1,132.55 feet/345.20 metres, SX 170785, a roundish, partly rocky, flat-topped peak, which rises gently and then fairly steadily from a tributary valley of the De Lank River to the west, gently from a col to the south-southeast, very gently and then gently from another one to the north-northwest and variably from the Fowey Valley to the east. OS Maps shows its highest point to be at, or close to, the OS 1:25000 map double field boundary, approximately 180 yards west-southwest of the 346-metre spot height. Two prehistoric cairns are situated on its summit area, with Catshole Tor on its west-southwestern edge. The Neolithic Long (chambered) Cairn is located on its southern flank and a prehistoric settlement is sited on its west-northwestern one.

15. Buttern Hill, 1,131.56 feet/344.90 metres, SX 174816, a somewhat elongated, partly rocky, roundish-topped peak, which rises fairly steeply and then gently from the Penpont Valley to the north-northwest, fairly steadily from one of its tributary dales to the east, gently from a col to the west and gently and then very gently from another to the south-southeast. OS Maps shows its highest point to be at the OS 1:25000 map 346-metre spot height, on the top of the second-most northerly of five Bronze Age round cairns marked on its summit area. A disused adit and a prehistoric settlement are situated on its north-northeastern flank and the source of the River Fowey is located at the foot of its west-southwestern one.

16. Newel Tor, 1,128.94 feet/344.10 metres, SX 236741, an elongated, partly rocky, flattish-topped peak, which rises fairly steeply and then gently from the Withey Valley to the east, quite gently and then gently from a dry valley to the north, quite gently and then very gently from a tributary dale of the River Fowey to the west and variably from another one to the south. OS Maps shows its rocky summit to be at the OS 1:25000 map 346-metre spot height. Its main tor rocks are situated just south and a little southeast of its highest point. There are quite a number of prehistoric remains on its slopes; Siblyback Moor and Carkeet Downs are located on its southwestern flank and Hill Tor and Hilltor Downs are sited on its west-southwestern one.

17. [Maiden Tor], 1,122.70 feet/342.20 metres, SX 158808, a small, partly rocky, flattish-topped peak on a north-northwest-projecting spur of Brown Willy, to the south. It rises quite gently from the De Lank Valley to the west, very gently and then quite gently from Roughtor March to the north-northwest and very gently from a depression to the east. OS Maps shows its summit to be at the west-northwestern edge of the OS 1:25000 map 342-metre spot height. However, the land drops by only 6.23 feet to a very high col linking it to its parent peak.

18. [Smallacoombe Downs], 1,111.22 feet/338.70 metres, SX 230747, an elongated, flat-topped peak, which rises gently and then very gently from a col to the west and variably from the Withey Valley to the east and one of its tributary dales to the north. OS Maps shows its highest point to be at the OS 1:25000 map 340-metre spot height. However, it fails to be an independent hill by only 8.33 feet and instead is an outlier of Newel Tor, to the south-southeast, to which it's linked by a high col. There are quite a number of prehistoric sites on its slopes, including a cairn close to its summit; almost all the hill is covered by coniferous plantations; Smallacoombe (rocky) Tor is situated at the east-northeastern end of its summit area and the remains of Trewortha medieval village are located on its lower north-northwestern flank.

19. Brown Gelly (Browngelly Downs), 1,109.25 feet/338.10 metres, SX 196726, an ovalish-shaped, flat-topped peak, which rises quite gently and then fairly steeply from a tributary valley of the River Fowey to the east, gently from Colliford Lake to the west, very gently and then quite gently from Dozmary Pool to the north and variably from a lake in china clay workings to the south. OS Maps surprisingly shows its summit to be at the OS 1:25000 map trig point, which is marked as standing at 338 metres, and not at the 342-metre spot height on the top of a prehistoric cairn, approximately 100 yards to the north-northwest, which is computed to be 0.98 feet lower. There are numerous prehistoric remains on its slopes; Whitebarrow Downs are situated on its west-southwestern flank; Colliford Downs are located on its southwestern one and Bunning's Park is sited on its west-southwestern one.

20. Kit Hill, 1,088.25 feet/331.70 metres, SX 374713, a round, isolated, flat-topped peak, which is the most easterly 1,000-foot hill in Cornwall. It rises steeply from the head of Mill Leat to the south, quite steeply and then steeply from cols to the west (at the village of Kelly Bray) and east and variably from a tributary valley of the River Tamar to the north. OS Maps shows its summit to be at the foot of an old mine chimney, in the middle of the OS 1:25000 map viewpoint symbol, marked as 334 metres, and a little west-northwest of the trig point. There are numerous disused quarries, shafts, mines and chimneys on its slopes, most of which are occupied by Kit Hill Country Park. Several prehistoric remains are sited on its slopes and Callington is situated on its southwestern flank. It is surrounded by major roads: the A 390 to the southeast and south, the A 388 to the southwest and west and the B 3257 to the northwest, north, northeast and east.

21. Ridge, 1,081.36 feet/329.60 metres, SX 242778, a somewhat elongated, slightly rocky, flat-topped peak, which rises fairly steeply and then gently from the Lynher Valley to the north and variably from the same dale to the east, the Withey Valley to the south and a col to the west. OS Maps shows its summit to be approximately in the middle of the OS 1:25000 map 330-metre contour ring. There are quite a number of prehistoric remains on its slopes, including the Nine Stones stone circle on its west-northwestern flank and Clitters Cairn on its northern one. Tolcarne Tor is situated on its north-northeastern flank and there is a fair amount of woodland on its lower eastern and east-southeastern slopes.

22. [Hill Tor (Hilltor Downs)], 1,077.76 feet/328.50 metres, SX 226739, a small, partly rocky peak on a west-southwest-projecting spur of Newel Tor. It rises quite gently from a tributary valley of the River Fowey to the west, gently from another to the south and very gently and then quite gently from a depression to the north. OS Maps shows its summit to be on the highest rock at the northeastern end of its main tor. However, the land drops by only 7.22 feet to a very high col linking it to its parent peak. A prehistoric field system is situated on its south-southwestern flank.

23. Leskernick Hill, 1,077.10 feet/328.30 metres, SX 183802, an oval-shaped, pretty rocky, flattish-topped peak, which rises fairly steadily and then gently from the head of a tributary dale of Penpont Water to the east, fairly steadily from the Fowey Valley to the west and variably from one of the latter's tributary dales to the south and the head of the Penpont Valley to the north. OS Maps shows its summit to be approximately 30 yards south-southeast of the OS 1:25000 map 329-metre spot height. There are numerous prehistoric remains on its slopes, including an extensive Bronze Age settlement on its upper south-southwestern flank.

24. Garrow Tor (Garrow Downs), 1,073.16 feet/327.10 metres, SX 144785, an elongated, partly rocky peak, which rises increasingly steeply from a col to the north and the De Lank River to the east, fairly steadily and then very gently from the latter to the south and very gently and then fairly gently from one of the tributary valleys of the same dale to the west. OS Maps shows its rocky tor summit to be just south-southwest of the OS 1:25000 map 330-metre spot height. There are a number of prehistoric remains on its slopes and a few small wooded areas on its lower ones.

CORNWALL'S 1,000-FOOT PEAKS 13

25. Hawk's Tor (Hawkstor Downs), 1,064.96 feet/324.60 metres, SX 253763, an elongated, very rocky peak, which rises gently from a tributary dale of the River Lynher to the south and variably from the Lynher Valley to the east, the Withey Valley to the north and a col to the west-southwest. OS Maps shows its rocky tor summit to be at the OS 1:25000 map 329-metre spot height. The fair discrepancy in the peak's altitude given by the online and paper versions of the OS's data might be explained by a limitation of the website's programme to calculate accurately the heights of rapidly-rising features. There are several prehistoric remains on its flanks, including Allabury Camp hillfort on its north-northeastern one, and there is a fair amount of woodland on its lower eastern and northern slopes.

26. Fox Tor, 1,053.81 feet/321.20 metres, SX 226785, an oval-shaped, fairly rocky peak, which rises very gently and then quite gently from a tributary valley of the River Lynher to the west, very gently and then gently from another to the east, very gently from a col to the south-southwest and variably from the Lynher Valley to the north. OS Maps shows its summit to be at the OS 1:25000 map trig point, which is marked as standing at 323 metres. There are a number of prehistoric remains on its slopes and some wooded areas on its lower ones. Redmoor Marsh is situated at the foot of its east-southeastern flank.

27. Carneglos Tor, 1,047.24 feet/319.20 metres, SX 202773, an elongated, flattish-topped peak, which rises quite gently from the Fowey Valley to the west, gently and then very gently from a col to the east and very gently from others to the south-southeast and north-northwest. OS Maps shows its summit to be approximately 85 yards north-northwest of the OS 1:25000 map 320-metre spot height. There are several prehistoric remains on its slopes, most of the upper ones of which, including its summit area, are wooded. The A 30 runs along the foot of its northern flank, broadly from south-southwest to north-northeast.

28. Trewortha Tor 1,036.75 feet/316.00 metres, SX 245758, a short, very rocky, east-northeast to west-running ridge, which rises increasingly steeply from the Withey Valley to the north, fairly steadily from the same dale and Trewortha Marsh to the west and very gently and then gently from cols to the east and south-southeast. OS Maps shows its rocky summit to be just southwest of the OS 1:25000 map 318-metre spot height. The unusual King Arthur's Bed rock is situated at the western end of its summit area; there are one or two prehistoric remains on its slopes and Tresellern Marsh is located at the foot of its west-northwestern flank.

29. [Codda Tor (Codda Downs)], 1,032.15 feet/314.60 metres, SX 176791, an elongated, slightly rocky, flattish-topped peak on a north-projecting spur of Catshole Downs, to the southwest. It rises very gently from a dry dale to the west and one of the tributaries of the River Fowey to the south, quite gently and then very gently from another of the latter to the north and variably from the Fowey Valley to the east. OS Maps shows its rocky summit to be at the southwestern edge of the OS 1:25000 map 318-metre spot height. However, it doesn't have a relative height of fifty feet and is linked to its parent peak by a very high col. Several prehistoric hut circles are situated on its eastern flank.

30. Butter's Tor (Butterstor Downs), 1,031.17 feet/314.30 metres, SX 154783, a somewhat elongated, partly rocky, roundish-topped peak, which rises fairly steadily from a tributary valley of the De Lank River to the east, quite gently from the same dale to the north, quite gently and then gently from the De Lank Valley to the west and quite gently and then very gently from one of its tributary streams to the south. OS Maps shows its summit to be just northwest of the OS 1:25000 map 316-metre spot height. A majority of its southern slopes are covered by woodland.

31. [Tregarrick Tor], 1,018.04 feet/310.30 metres, SX 241711, a small, partly rocky peak on a south-southwest to west-southwest-curving spur of Stowe's Hill, well over 1¼ miles to the east-northeast. It rises fairly steadily from Siblyback Lake to the west and one of its tributary valleys to the south and very gently from the head of another tributary dale to the north. OS Maps shows its summit to be at the southern end of the OS 1:25000 map 310-metre very small contour ring. However, it doesn't have a relative height of fifty feet and is linked to its parent peak by a very high col. A prehistoric hillfort, a round cairn and a tor cairn are all situated on its summit area and several other prehistoric remains are located on its slopes.

32. Louden Hill, 1,017.39 feet/310.10 metres, SX 137803, a somewhat elongated, partly rocky peak, which rises quite gently from Stannon (china clay) Works to the west, quite gently and then very gently from a tributary valley of the River Camel to the north, very gently and then steadily from a col to the east-northeast and variably from a tributary dale of the De Lank River to the south. OS Maps shows its summit to be at the western edge of the OS 1:25000 map 315-metre spot height. There is a fair discrepancy in the peak's altitude given by the online and paper versions of the OS's data, but there is no obvious explanation for it. There are quite a number of prehistoric remains on its slopes and Steping Hill is situated on its southern flank.

33. Priddacombe Downs, 1,014.76 feet/309.30 metres, SX 162771, a slightly elongated, flattish-topped peak, which rises variably from a tributary valley of the De Lank River to the north, quite gently and then gently from the same dale to the west and gently and then very gently from another one to the south and east. OS Maps shows its summit to be at the north-northeastern edge of the OS 1:25000 map spot height. Most of its western slopes are covered by woodland and there are smaller woods on its southern and south-southwestern ones. A Bronze Age platform cairn is situated at its summit and the Priddacombe Downs Nature Reserve covers part of its slopes.

34. [Whitebarrow Downs], 1,007.55 feet/307.10 metres, SX 189722, a small, partly rocky peak on a west-southwest to south-southwest-curving spur of Brown Gelly, to the east-northeast. It rises gently from Colliford Lake (via Bunning's Park) to the west, very gently from one of its tributary valleys to the north and variably from one of the River Fowey to the south. OS Maps shows its summit to be at the northern edge of the OS 1:25000 map prehistoric earthwork within the 300-metre contour bulge. However, the land drops by only 11.16 feet to a very high col linking it to its parent peak. Colliford Downs are situated on its south-southwestern flank and an unnamed lake is located at the foot of its south-southeastern one.

35. Lowlands Down, 1,006.89 feet/306.90 metres, SX 210749/209750, an elongated, twin-summited, flattish-topped peak, which rises very gently from a col to the north-northwest, variably from another to the east and the Fowey Valley to the south and fairly steadily and then very gently from the same dale to the west. OS Maps shows its two summits, of identical height, to be at the OS 1:25000 map 307-metre spot height and on top of a prehistoric cairn, approximately 150 yards to the northwest. A restored stone circle is situated towards the northwestern end of its summit area; Smith's Moor lies beyond it; there are a number of prehistoric remains on its slopes; there is a considerable amount of woodland on its northern ones; Goodaver Downs are located on its north-northeastern flank and the disused Goodaver Mine is sited on its southern one.

36. Hendraburnick Down, 1,004.59 feet/306.20 metres, SX 147884, an elongated peak, which is the most northerly 1,000-foot hill in Cornwall. It rises quite gently and then gently from a tributary dale of the River Valency to the west and variably from the Inny Valley to the south, the Ottery Valley to the east and the Valency Valley to the north. OS Maps shows its highest point to be on the top of its summit area's southeastern Bronze Age barrow (marked on the OS 1:25000 map as standing at 309 metres), which is just 0.32 feet higher than that computed of the foot of the trig point situated on the nearby northwestern barrow. A radio station, several telecommunications masts and Tich (Bronze Age) Barrow are located on its greater summit area; there are quite a number of other prehistoric remains on its flanks and the sources of the River Camel, the River Valency, the River Ottery and the River Inny are all sited on its slopes. The A 39 runs across the southern and eastern sides of its greater summit area, broadly from south-southwest to north-northeast.

37. Hensbarrow Beacon (Hensbarrow Downs), 1,003.94 feet/306.00 metres, SW 997575, the remains of a substantial, flattish-topped hill, which is the most southerly and westerly 1,000-foot peak in Cornwall. It is almost surrounded and has largely been destroyed or disfigured by a whole series of china clay works, which have produced tips higher than its natural peak! It rises quite steeply from the head of an unnamed valley to the north, fairly steadily from Vale Pleasant to the south and quite gently from a col to the west. OS Maps shows its summit to be approximately 50 yards northeast of the OS 1:25000 map trig point, although other evidence indicates that it is situated on the top of Hensbarrow Bronze Age round cairn (later used as a beacon), at a spot quite close to the pillar, which is marked as standing at 312 metres. There is a sizable discrepancy in the peak's altitude given by the online and paper versions of the OS's data, but there is no obvious explanation for it. Several telecommunications masts are situated on its upper slopes and the B 3274 runs across its eastern and northern ones, broadly from south-southeast to north-northwest.

38. (Brockabarrow Hill), 1,002.62 feet/305.60 metres, SX 160747, an elongated, flattish-topped peak, which rises fairly gently from the Warleggan Valley to the west and a tributary dale of the De Lank River to the east and variably from Colliford Lake to the south-southeast and the De Lank Valley to the north-northwest. OS Maps shows its rocky summit to be on the top of a prehistoric cairn towards the southern end of the OS 1:25000 map 300-metre elongated contour ring. The hill is unnamed on the OS 1:25000 and 1:50000 maps, but it seems appropriate to name it after Brockabarrow Common, which is situated on its slopes. There are quite a number of prehistoric remains on its flanks; Blacktor Downs are located on its south-southwestern one and the A 30 runs across its southern one, broadly from east to west.

39. (Condolden Hill), 1,000.00 feet/304.80 metres, SX 090872, a roundish, flat-topped peak, which rises quite steeply and then fairly gently from the Rocky Valley to the north, fairly steadily and then very gently from an unnamed valley to the south and variably from the Atlantic Ocean to the west and the Camel Valley to the east. OS Maps shows its summit to be approximately 40 yards north of the OS 1:25000 map trig point, which is marked as standing at 308 metres, but other evidence indicates its highest point to be by the pillar on the top of the prehistoric Condolden (bowl) Barrow. Although the hill is unnamed on the OS 1:25000 and 1:50000 maps, it seems appropriate to name it after the barrow and Condolden Farm, which is situated on its south-southeastern flank. The village of Tintagel is located on its lower west-northwestern flank and the hamlet of Trewarmett on its west-southwestern one. The B 3263 runs along the foot of its southern slopes, broadly from east-southeast to west-northwest, and the B 3266 crosses its eastern ones, essentially from south-southwest to north-northeast.

40. Hawk's Tor (Hawkstor Downs), 989.50 feet/301.60 metres, SX 141755, a roundish peak, which rises gently from the head of a tributary valley of the De Lank River to the south, fairly gently from the same dale to the west, very gently and then fairly gently from another such tributary valley to the north and variably from the head of the Warleggan Valley to the east. OS Maps shows its rocky summit to be at the OS 1:25000 map trig point, which is marked as standing at 307 metres. However, it doesn't qualify as a 1,000-foot hill according to OS Maps, which computes its height as being 5.40 metres lower than that marked on the map! The fair discrepancy in the peak's altitude given by the online and paper versions of the OS's data might be explained by a limitation of the website's programme to calculate accurately the heights of rapidly-rising features. The Stripple Stones Neolithic henge and stone circle are situated on its south-southeastern flank; Carkees Down and Tor are located on its north-northwestern one and the Carkees Down Long House ancient settlement is sited on its west-northwestern one.

4 Devon's 1,000-Foot Peaks

Devon is largely an undulating county, but its south central area is dominated by the great granite Dartmoor upland, on and around which most of its 1,000-foot peaks are located. Two of the hills at its northern end (High Willhays and Yes Tor) reach over 2,000 feet, making Devon's tallest 1,000-foot peak (Hangingstone Hill) only the third highest in the shire. An important feature of Dartmoor is its rocky tors, a large number of which qualify as 1,000-foot peaks. In addition, in the very north of the county, there is another notable upland area, stretching inland from the Bristol Channel, which includes the sedimentary 1,000-foot hills of western Exmoor.

1. Hangingstone Hill, 1,978.35 feet/603.00 metres, SX 617861, an elongated, partly rocky, flat-topped peak, the third-highest in Devon, which shares a small plateau with nearby Whitehorse Hill, to the south. It rises fairly steeply and then gently from the infant Taw Valley to the west, very gently and then quite gently from the Steeperton Valley to the north, very gently from a col to the south and variably from the upper Walla Valley to the east. OS Maps shows its summit to be on the top of a Bronze Age round cairn, at the OS 1:25000 map 603-metre spot height. Wild Tor is situated on its north-northeastern flank and Taw Head at the foot of its west-southwestern one.

2. [Whitehorse Hill], 1,973.75 feet/601.60 metres, SX 616854, a flat-topped peak, which shares a small plateau with, and is an outlier of, nearby Hangingstone Hill, to the north. It rises gently and then very gently from East Dart Head to the west and Teign Head to the south and variably from the North Teign Valley to the east. OS Maps shows its summit to be just to the southwest of the OS 1:25000 map 602-metre spot height, but the land drops by only 6.89 feet to a very high col linking it to its parent peak. Approximately 50 yards to the northeast of its summit is an important Bronze Age cist, which was discovered in 2001; Quintin's Man prehistoric cairn is situated on its south-southeastern flank; the Great Varracombe Valley is located at the foot of its southeastern one; Walla Brook Head is sited at the bottom of its east-northeastern one and Manga Hill is found on its east-southeastern one.

3. Cut Hill, 1,972.44 feet/601.20 metres, SX 598827, a flat-topped peak, which rises fairly steeply and then gently from the head of Cut Comb Water to the west and the Cut Hill Valley to the east, very gently and then gently from a col to the south and variably from another to the north-northeast. OS Maps shows its highest point to be approximately 35 yards southwest of the OS 1:25000 map 603-metre spot height. The hill is a significant watershed, with Tavy Head, West Dart Head and the sources of Cut Combe Water, Cut Hill Water and the Cut Combe Stream all on its flanks. A Neolithic stone row, discovered in 2004, lies on its summit area and Fur Tor is situated on its west-northwestern flank.

4. Amicombe Hill, 1,917.65 feet/584.50 metres, SX 567872, an elongated, partly rocky, flat-topped peak, which rises steeply and then gently from the West Okement Valley to the north and east and very gently from the Tavy Valley to the south and the Rattle Valley to the west. OS Maps shows its summit to be at the OS 1:25000 map 584-metre spot height. Kitty Tor is situated on its northern shoulder; Steng-a-Tor is located on its northern flank; Green Tor is sited on its south-southwestern one and Woodcock Hill, Hunt Tor and Gren Tor are found on its west-northwestern one.

5. Black Hill, 1,911.75 feet/582.70 metres, SX 604846, an ovalish-shaped, flat-topped peak, which rises quite gently and then gently from the Black Ridge Valley to the west, very gently from a col to the south-southwest and East Okement Head (via Okement Hill) to the north and variably from the East Dart Valley to the east. OS Maps shows its summit to be at the OS 1:25000 map 584-metre spot height. Little Kneeset is situated on its west-southwestern flank, Black Ridge on its northwestern one, West Okement Head and Cranmere Pool on its northern one and East Dart Head at the foot of its north-northeastern one.

6. Great Links Tor, 1,897.97 feet/578.50 metres, SX 550867, an elongated, partly rocky peak, which rises quite steadily and then fairly steeply from the Lyd Valley to the west, fairly steadily from a dry valley to the north, quite gently from the head of the Doetor Valley to the south and variably from the Rattle Valley to the east. OS Maps shows its summit to be at the OS 1:25000 map trig point, which is marked as standing at 586 metres, but other evidence indicates it to be on the top of an impressive large rock tor just to the east of it. The sizable discrepancy in the peak's altitude given by the online and paper versions of the OS's data might be explained by a limitation of the website's programme to calculate accurately the heights of rapidly-rising features. There are several prehistoric remains on its slopes; Higher and Lower Dunna Goat crags are situated on its east-southeastern flank; Little Links Tor is located on its west-northwestern one; Arms Tor is sited on its west-southwestern one and Brat Tor and Widgery Cross can be found on its southwestern one.

7. [Fur Tor], 1,876.31 feet/571.90 metres, SX 587830, a rocky peak on a north-northwest-projecting spur of Cut Hill, to the east-southeast. It rises very gently and then fairly steeply from the Tavy Valley to the west and variably from the Cut Combe Valley to the east and north. OS Maps shows its highest point to be on its summit tor at the OS 1:25000 map 572-metre spot height. However, the hill doesn't have a relative height of fifty feet and is linked by a high col to its parent peak. Pinswell is situated on its west-northwestern flank.

8. [Black Ridge], 1,873.36 feet/571.00 metres, SX 596855, a partly rocky, flat-topped peak on a west-projecting spur of Black Hill, to the south-southeast. It rises quite steadily and then very gently from the Black Ridge Valley to the south, quite gently and then very gently from the West Okement Valley to the north and variably from the head of a tributary dale of the former to the west. OS Maps shows its summit to be approximately 100 yards east of the OS 1:25000 map 573-metre spot height, but the land drops by only 12.14 feet to a very high col linking it to its parent peak. West Okement Head and Cranmere Pool are situated on its east-northeastern flank and Great Kneeset is located on its west-northwestern one.

9. [Woodcock Hill], 1,866.14 feet/568.80 metres, SX 561879, a small peak on a west to south-southwest-curving spur of Amicombe Hill, to the east-southeast. It rises steeply and then very gently from the West Okement Valley to the north, fairly steadily and then very gently from the Lyd Valley to the west and very gently from the head of the Rattle Valley to the south-southwest. OS Maps shows its summit to be roughly in the middle of the OS 1:25000 map 560-metre contour bulge, but the land drops by only 10.17 feet to a very high col linking it to its parent peak. Hunt Tor is situated on its west-southwestern flank, Gren Tor on its western one and Lyd Head on its northwestern one.

10. [Okement Hill], 1,855.64 feet/565.60 metres, SX 602877, a slightly rocky, flat-topped peak on a north-projecting spur of Black Hill, nearly two miles to the south. It rises quite steeply and then gently from East Okement Head to the north-northeast, very gently from the Brim Valley to the west and variably from the Taw Valley to the east. OS Maps shows its summit to be approximately 65 yards east of the OS 1:25000 map 564-metre spot height. However, the hill doesn't have a relative height of fifty feet and is linked to its parent peak by a very high col. Vergyland Combe is situated on its south-southwestern flank.

11. [Great Kneeset], 1,851.71 feet/564.40 metres, SX 589859, a partly rocky, flat-topped peak on a west-northwest-projecting spur of Black Hill (and, more immediately, Black Ridge), approaching 1¼ miles to the southeast. It rises quite gently from a tributary dale of the Black Ridge Brook to the south and variably from the West Okement Valley to the north and a col to the west. OS Maps surprisingly shows its highest point to be approximately 60 yards east-northeast of the OS 1:25000 map 567-metre spot height on its summit tor. However, the hill doesn't have a relative height of fifty feet and is separated from its parent peak only by a very high col. Jackman's Bottom is situated at the foot of its northeastern flank.

12. Cosdon Hill (Cosdon Beacon), 1,798.56 feet/548.20 metres, SX 635914, a bulky, elongated, partly rocky, round-topped peak, which rises steeply from the Taw Valley to the north, variably from the same dale to the west, quite steeply and then fairly steadily from the head of Cheriton Combe to the east and gently and then fairly steadily from a col to the south-southwest. OS Maps shows its summit to be approximately 50 yards southwest of the OS 1:25000 map trig point (which is marked as standing at a height of 550 metres), although other evidence indicates it to be just northwest of the pillar. There are numerous prehistoric remains on its slopes, including a cist, with three parallel stone rows running downhill from it, on its eastern flank. South Tawton Common is situated on its southern flank, White Hill on its west-southwestern one, Skaigh Warren on its northern one, Foxes' Holt on its north-northeastern one and Raybarrow Pool at the foot of its south-southeastern one.

13. [Devil's Tor], 1,793.64 feet/546.70 metres, SX 597797, a small, rocky-topped peak on a south-projecting spur of Cut Hill, approaching two miles to the north. It rises quite steeply and then gently from the Cowsic Valley to the west, gently and then very gently from a col to the south-southeast and very gently from a dry valley to the east. OS Maps shows its summit to be approximately 150 yards northeast of the OS 1:25000 map 549-metre spot height, but the land drops only by 12.80 feet to a very high col linking it to its parent peak. A tall Bronze Age standing stone, the Beardown Man, is situated to the west of the spot height and Lydford Tor is located on its southern flank.

14. [Wild Tor], 1,792.98 feet/546.50 metres, SX 621871, a small, partly rocky, flat-topped peak on a north-northeast-projecting spur of Hangingstone Hill. It rises increasingly gently from the Walla Valley to the east, gently from the Steeperton Valley to the west and variably from a col to the north-northeast. OS Maps shows its summit to be approximately 60 yards southwest of the OS 1:25000 map 547-metre spot height. However, the land drops by only 11.16 feet to a very high col linking it to its parent peak. Its main tor is situated on its upper northern flank, a tinner's hut on its north-northwestern one, a prehistoric cairn on its east-northeastern one and Gallaven Mire at the foot of its northeastern one.

15. [Rough Tor], 1,792.65 feet/546.40 metres, SX 606798, a small, rocky-topped peak on an east-projecting spur of Cut Hill (and, more immediately, Devil's Tor), well over 1¾ miles to the north-northwest. It rises fairly steadily and then quite gently from the West Dart Valley to the east, fairly steadily and then very gently from the same dale to the south and very gently from the Summer Valley to the north. OS Maps shows its highest point to be at the OS 1:25000 map 547-metre spot height on its fairly flat summit tor, which is partly occupied by small military buildings. However, it fails by just 1.12 feet to be an independent hill and is linked to its parent peak by a high col. There are one or two prehistoric remains on its slopes and Crow Tor is situated on its southern flank.

16. [Watern Tor], 1,787.07 feet/544.70 metres, SX 628861, an elongated, flat-topped, north-projecting spur of Hangingstone Hill (and more immediately Whitehorse Hill), to the west. It rises gently and then quite steeply from the North Teign Valley to the east, gently from the Walla Valley to the west and variably from the latter to the north. OS Maps shows its summit to be at the OS 1:25000 map 545-metre spot height. However, the hill doesn't have a relative height of fifty feet and is linked to its parent peak by a high col to the south. The tor itself is more than 700 yards to the north of the hill summit, with the strangely eroded Thirlstone a little further on. There are one or two prehistoric remains on its slopes; Hew Down is situated on its eastern flank; Hawthorn Clitter is located on its east-southeastern one and Walla Brook Head is sited at the foot of its south-southwestern one.

17. [Maiden Hill], 1,780.84 feet/542.80 metres, SX 588795, a somewhat elongated, partly rocky, flat-topped peak on a south-projecting spur of Cut Hill, more than 2¼ miles to the north-northeast. It rises steeply and then gently from the Cowsic Valley to the east, fairly steadily and then gently from Conies Down Water to the south and gently from the Walkham Valley to the west. OS Maps shows its summit to be approximately in the middle of the OS 1:25000 map 540-metre contour ring. However, the hill doesn't have a relative height of fifty feet and is linked to its parent peak by a very high col. Its whole area is situated within Merrivale (firing) Range; Conies Down Tor is located on its southern flank; Conies Down Neolithic Double Stone Row is sited on its south-southwestern one; Conies Down is found on its west-southwestern one and a series of prehistoric stone hut circles lie on its southeastern one.

18. [Flat Tor], 1,775.26 feet/541.10 metres, SX 607817, a small, flat-topped peak on an east to south-southeast-curving spur of Cut Hill, approaching a mile to the northwest. It rises fairly steadily and then very gently from Cut Hill Water to the north and very gently from the West Dart Valley to the south and the East Dart Valley to the east. OS Maps shows its summit to be approximately in the middle of the OS 1:25000 map 540-metre contour bulge, just over 200 yards north-northwest of the 542-metre spot height, but the land drops by only 9.18 feet to a very high col linking it to its parent peak.

19. Broad Down, 1,769.03 feet/539.20 metres, SX 614804, a slightly rocky, roundish-topped peak on a small plateau. It rises quite gently and then gently from the West Dart Valley to the west, gently from one of its tributary dales to the south, very gently from one of the East Dart River to the north and variably from the East Dart Valley to the east. OS Maps shows its summit to be at the OS 1:25000 map 539-metre spot height. There are one or two prehistoric remains on its slopes.

20. Rattlebrook Hill (Chat Tor), 1,768.04 feet/538.90 metres, SX 555852, an ovalish-shaped, partly rocky, flat-topped peak, which rises steeply and then gently from the Rattle Valley to the east, quite steeply and then very gently from the Doetor Valley to the west and variably from the Tavy Valley to the south. The hill lies south of Great Links Tor, from which it's separated by a high col, and it qualifies as an independent peak by just 0.19 feet! OS Maps shows its highest point to be around the OS 1:25000 map 542-metre spot height, which is surprisingly marked just north of Chat Tor, the summit outcrop. There are several prehistoric remains on its slopes and Sharp Tor, Doe Tor and Wallabrook Head are situated on its west-southwestern flank.

21. [Winney's Down], 1,766.08 feet/538.30 metres, SX 621825, an elongated, flat-topped peak, which rises gently from a tributary dale of the North Teign River to the north, fairly gently and then very gently from the East Dart Valley to the west and variably from the latter to the south. OS Maps shows its summit to be approximately 50 yards southwest of the OS 1:25000 map 539-metre spot height. However, it fails to be an independent peak by just 5.70 feet and instead is an outlier of Hangingstone Hill (and, more immediately, Whitehorse Hill) more than 1¾ miles to the north, to which it's linked by a high col. Statts House (a peat cutters' ruined shelter) is situated on its summit area; Kit Rocks are located on its western flank and Little Varracombe is sited at the foot of its northeastern one.

22. [Corn Ridge], 1,760.17 feet/536.50 metres, SX 552891, a small, flattish-topped peak on a northwest-projecting spur of Amicombe Hill, more than

1½ miles to the south-southeast. It rises steeply from a tributary valley of the West Okement River to the north, fairly steeply and then quite gently from one of the Crandford Brook (via Lake Down) to the west and variably from the West Okement Valley to the east. OS Maps shows its summit to be on the top of a rocky outcrop approximately 350 yards north of the OS 1:25000 map 537-metre spot height. However, the hill doesn't have a relative height of fifty feet and is linked to its parent peak by a high col. The impressive Branscombe's Loaf rocky tor and a Bronze Age cairn cemetery are situated on its summit area; Coombe Down is located on its west-southwestern flank and Shelstone Tor is sited on its north-northeastern one.

23. Sittaford Tor, 1,757.87 feet/535.80 metres, SX 633830, an elongated, pretty rocky peak, which rises fairly steadily from the North Teign Valley to the north, gently from one of its tributary dales to the west, very gently from the Winney's Down Valley to the south-southwest and variably from the South Teign Valley to the east. OS Maps shows its summit to be approximately just south-southeast of the OS 1:25000 map field boundaries' junction within the 530-metre contour ring, at or near the unmarked 538-metre spot height. Sittaford Stone Circle, which was discovered in 2007, is situated about 275 yards southwest of its summit; a series of stone hut circles are located on its southeastern flank; the source of the South Teign River lies at the foot of its east-southeastern one and the Grey Wethers restored double stone circle is sited on its eastern one.

24. [West Mill Tor], 1,746.06 feet/532.20 metres, SX 587909, an elongated, rocky, north-projecting spur of High Willhays, more than a mile to the south-southwest. It rises quite steeply from the Moor Valley to the north, fairly steadily and then quite steeply from the Red-a-ven Valley to the west and variably from the Black-a-ven Valley to the east. OS Maps shows its summit to be at the OS 1:25000 map 541-metre spot height. The large discrepancy in the peak's altitude given by the online and paper versions of the OS's data might be explained by a limitation of the website's programme to calculate accurately the heights of rapidly-rising features. The hill doesn't have a relative height of fifty feet and is linked to its parent peak by a high col.

25. Great Mis Tor, 1,742.78 feet/531.20 metres, SX 562769, a roundish, rocky peak, which rises steadily and then quite steeply from the Walkham Valley to the west, variably from the same dale to the north and one of its tributary valleys to the south and very gently and then gently from the Blackbrook Valley to the east. OS Maps shows its highest point to be on the top of its summit tor, approximately 50 yards northwest of the OS 1:25000 538-metre spot height. The considerable discrepancy in the peak's altitude given by the online and paper versions of the OS's data might be explained by a limitation of the website's programme to calculate accurately the heights of rapidly-rising features. The Mistor Pan rock basin is situated on its summit area; Little Mis Tor and a series of Bronze Age stone hut circles are located on its southern flank; a whole range of prehistoric and historic remains are sited on its west-southwestern to south-southwestern ones, most especially Merrivale Stone Rows; Greena Ball can be seen on its north-northeastern one and Mistor Marsh lies on its northeastern one.

26. Hamel Down, 1,741.47 feet/530.80 metres, SX 705799, the highest part of a small, elongated, north-south plateau, which rises quite steeply from the West Webburn Valley to the west and variably from the same dale to the south (via Hameldown Beacon), its head to the north (via Hameldown Tor) and the East Webburn Valley to the east. OS Maps shows its summit to be just north of the OS 1:25000 map 532-metre spot height. The remains of Hamel Down (ancient) Cross are situated approximately 275 yards north-northwest of the hill top; a series of Bronze Age round cairns (including Broad Barrow, Single Barrow and Two Barrows) stretch south for two-thirds of a mile along the plateau from its summit; Berry Pound prehistoric enclosure is located on its east-northeastern flank; a number of other prehistoric remains lie on its slopes and Blackaton Down is sited on its south-southwestern flank.

27=. [Hameldown Tor], 1,732.94 feet/528.20 metres, SX 703805, a partly rocky, north-northwest-projecting spur of Hamel Down, which rises quite steeply from the West Webburn Valley to the west, quite steeply and then gently from a tributary dale of the River Bovey to the east and variably from the head of the West Webburn Valley to the north. OS Maps shows its highest point to be on its summit Bronze Age ring cairn, by the OS 1:25000 map trig point, which stands at 529 metres. However, the hill doesn't have a relative height of fifty feet and is linked to its parent peak by a high col. The source of the East Webburn River is situated on its east-southeastern flank; its lower eastern one is wooded and the impressive Bronze Age village of Grimspound is located at the foot of its north-northwestern one.

27=. Hare Tor, 1,732.94 feet/528.20 metres, SX 550842, an impressive rocky peak, which rises fairly steadily and then gently from Dead Lake to the east, gently and then fairly steadily from the Walla Valley to the west, very gently and then fairly steadily from a col to the north-northeast and variably from the Tavy Valley to the south. OS Maps shows its summit to be at the OS 1:25000 map 531-metre spot height, on the top of its rocky tor. Its whole area is part of Willsworthy (firing) Range; there are quite a number of prehistoric remains on its slopes and Ger Tor, Nattor Down and Nat Tor are situated on its south-southwestern flank.

29. Steeperton Tor, 1,726.38 feet/526.20 metres, SX 618887, an elongated, rocky-topped peak, which rises quite steeply from the Steeperton Gorge to the west and north, fairly steadily from the Steeperton Valley to the east and very gently and then gently from a col to the south-southwest. OS Maps shows its summit to be on the top of a conical tor, on the northwestern edge of the OS 1:25000 map 532-metre spot height. The fair discrepancy in the peak's altitude given by the online and paper versions of the OS's data might be explained by a limitation of the website's programme to calculate accurately the heights of rapidly-rising features. Artillery ranges are situated on its slopes; the disused Knack Mine is located on its southwestern flank; a Bronze Age stone hut circle settlement lies on its northeastern one and a tinner's blowing house is sited on its south-southeastern one.

30. Higher White Tor, 1,720.80 feet/524.50 metres, SX 619785, a wedge-shaped, pretty rocky peak, which rises quite steeply from Hollowcombe Bottom to the east, increasingly gently from the West Dart Valley to the west, very gently and then gently from a col to the south-southwest and very gently from another to the north-northwest. OS Maps shows its highest point to be at the OS 1:25000 map 527-metre spot height, which is situated on its summit tor. The remnants of a Neolithic double stone row are situated on its south-southwestern flank; other prehistoric remains are located on its southeastern one and Lower White Tor is sited on its northern one.

31. [Sharp Tor], 1,698.16 feet/517.60 metres, SX 550848, a small, rocky peak on a west-southwest-projecting spur of Rattlebrook Hill. It rises fairly steeply and then very gently from the Rattle Valley to the east, gently from a depression to the south and variably from the Lyd Valley (via Doe Tor and Doetor Common) to the west. OS Maps shows its summit to be just west of the number "5" in the unmarked OS 1:25000 map 519-metre spot height, but the land drops by only 7.54 feet to a very high col linking it to its parent peak. Its main rocky tor is approximately 100 yards to the northwest of its summit. Its whole area is a part of Willsworthy (firing) Range and Wallabrook Head is situated on its west-southwestern flank.

32. [Lynch Tor], 1,696.52 feet/517.10 metres, SX 566805, an elongated, pretty rocky, flattish-topped peak on a south-southwest-projecting spur of Cut Hill (and, more immediately, Walkham Head), well over 2¼ miles to the east-northeast. It rises fairly steadily and then quite steeply from the head of the Baggator Valley to the west and very gently from the Walkham Valley to the east and a col to the south. OS Maps shows its summit to be just to the east of the OS 1:25000 map 517-metre spot height. However, it fails by only 7.45 feet to be an independent hill and is linked to its parent peak by a high col. The Bronze Age Limsboro Cairn is situated on its summit area and Wapsworthy Common is located on its southwestern flank.

33. Ryder's Hill, 1,688.32 feet/514.60 metres, SX 659691, an elongated, flat-topped peak, which rises fairly steadily and then quite gently from the Mardle Valley to the east, fairly steadily and then very gently from the O Valley to the north, gently from the Avon Valley to the west and very gently from a col to the south. OS Maps computes its highest point to be approximately 60 yards north of the OS 1:25000 map trig point, which is marked as standing at 515 metres. Beneath the pillar are the remains of a Bronze Age cairn and Petre's Bound Stone (a boundary marker) is very close by. Mardle Head is situated on its eastern flank, Holne Ridge is located on its north-northeastern one and Avon Head and Avon Head Mires are sited at the foot of its west-northwestern one.

34. [Little Kneeset], 1,687.66 feet/514.40 metres, SX 591842, a small peak on a flat-topped, west-projecting spur of Black Hill, to the east-northeast. It rises fairly steadily and then very gently from the Black Ridge Valley to the west and quite gently from the same dale to the north and Cut Combe Water to the south. OS Maps shows its summit to be just southwest of the OS 1:25000 map 515-metre spot height, but the land drops by only 12.46 feet to a very high col linking it to its parent peak. A peat pass runs from north to south just west of its summit.

35. [Conies Down], 1,683.73/513.20 metres, SX 581792, a small, flat-topped peak on a west-projecting spur of Cut Hill (and, more immediately, Maiden Hill), approaching 2½ miles to the north-northeast. It rises quite gently from the Walkham Valley to the west, gently from one of its tributary dales to the north and very gently from a col to the south. OS Maps shows its summit to be approximately in the middle of the OS 1:25000 map 510-metre contour bulge. However, the land drops by only 5.58 feet to a very high col linking it to its parent peak. It is entirely situated inside the Merrivale (firing) Range.

36. North Hessary Tor, 1,674.87 feet/510.50 metres, SX 578742, a large, partly rocky, flat-topped peak, which rises quite gently from the Meavy Valley to the south and variably from the Blackbrook Valley to the east and north and the Walkham Valley to the west. OS Maps shows its highest point to be on its small summit tor, at the OS 1:25000 map trig point, which is marked as standing at 517 metres. The considerable discrepancy in the peak's altitude given by the online and paper versions of the OS's data might be explained by a limitation of the website's programme to calculate accurately the heights of rapidly-rising features. A television and radio transmitter station is situated on its summit area; there are quite a number of prehistoric remains on its slopes; Meavy Head lies on its south-southeastern flank; the village of Princetown is located on its east-southeastern one; Rundlestone Tor is sited on its north-northwestern one; Hollow Tor can be seen on its west-northwestern one and the disused Foggintor Quarries are found on its west-southwestern one. The B 3212 runs across its southern flank, broadly from west-southwest to east-northeast, and the B 3357 curves across its eastern and northern ones, essentially from east-southeast to west-northwest.

37. East Mill Tor, 1,674.54 feet/510.40 metres, SX 599898, an elongated, rocky peak, which rises fairly steadily from the Black-a-ven Valley to the west, variably from the same dale to the north, gently and then steeply from the East Okement Valley to the east and very gently from a col to the south. OS Maps shows its summit to be at the OS 1:25000 map 513-metre spot height.

38. Beardown Tors, 1,673.23 feet/510.00 metres, SX 602773, an elongated, rocky, triple-summited peak, which rises fairly steeply from the West Dart Valley to the east, quite steadily from the Cowsic Valley to the west, fairly steadily and then very gently from the latter to the south (via Beardown Hill) and variably from a col to the north. OS Maps shows its main summit to be on the top of its most westerly tor, indicated on the OS 1:25000 map to be 513 metres high. The summit of its second-highest tor is approximately 300 yards to the east-northeast and stands at 1,670.28 feet, according to OS Maps, whilst the most northerly tor (about 400 yards north of there) is computed to reach an altitude of 1,648.62 metres. Several Bronze Age stone hut circles are located on its southeastern flank; there are one or two other prehistoric remains on its slopes and an area of woodland is situated on its lower southern ones.

39. White Ridge, 1,657.48 feet/505.20 metres, SX 648821, an elongated, flattish-topped peak, which rises quite steeply and then very gently from the Assycombe Valley to the east, fairly steadily and then very gently from a tributary dale of the East Dart Valley to the west and variably from the Stannon Valley to the south and the South Teign Valley to the north. OS Maps shows its summit to be at the OS 1:25000 map 506-metre spot height. There are a series of prehistoric remains on its slopes; Stannon Tor is situated on its south-southwestern flank; most of its eastern and northeastern ones are covered by Fernworthy Forest and the source of the South Teign River is located at the foot of its northwestern one.

40. Longaford Tor, 1,645.34 feet/501.50 metres, SX 615779, a small, somewhat elongated, rocky peak, which rises fairly steadily and then steeply from the West Dart Valley to the west, very gently and then steeply from a col to the north and variably from the Muddilake Valley (via Crockern Tor and Littaford Tors) to the south and the Cherry Valley to the east. OS Maps shows its highest point to be at the OS 1:25000 507-metre spot height, which is situated at the top of its impressive pyramidal summit tor. There is a fair discrepancy in the peak's altitude given by the online and paper versions of the OS's data, but there is no obvious explanation for it. There are a number of prehistoric remains and pillow mounds on its slopes and Wistman's Wood National Nature Reserve is located on its south-southwestern flank.

41. [Cocks Hill], 1,641.73 feet/500.40 metres, SX 568789, a small, flattish-topped, south-southeast-projecting spur of Cut Hill (and, more immediately, Lynch Tor and Walkham Head), just over three miles to the north-northeast. It rises fairly steadily and then gently from the Walkham Valley to the east, increasingly gently from the same dale to the south and very gently from a col to the west. OS Maps shows its summit to be at the OS 1:25000 map 501-metre spot height. However, it doesn't have a relative height of fifty feet and is linked to its parent peak by a high col. There are several prehistoric remains on its slopes, including White Barrow on its northern flank, and Petertavy Great Common and Langstone Moor are situated on its west-southwestern slopes.

42. Assycombe Hill, 1,631.89 feet/497.40 metres, SX 664820, an elongated, roundish-topped peak, which rises quite steeply from the Assycombe Valley to the west, fairly steadily and then gently from the North Walla Valley to the east and variably from Fernworthy Reservoir to the north and a col to the south-southwest. OS Maps shows its summit to be just east-southeast of the OS 1:25000 map 497-metre spot height. The majority of the hill, including most of its summit area, is situated in Fernworthy Forest; there are quite a number of prehistoric remains on its slopes, including a double stone row, and Hurston Ridge is located on its eastern flank.

43. Hookney Tor (Hookney Down), 1,625.00 feet/495.30 metres, SX 698812, a roundish peak, which rises fairly steadily from the West Webburn Valley to the south, gently and then fairly steadily from one of its headwater streams to the west, gently from a col to the north and very gently from another to the east. OS

Maps shows its highest point to be just south of its summit prehistoric cairn, although other evidence indicates the cairn itself to be its highest point. A number of rocky outcrops are situated on its summit area and a series of prehistoric stone hut circles on its north-northeastern flank.

44. [Snowdon], 1,618.77 feet/493.40 metres, SX 668683, an elongated, flat-topped peak on a south-projecting spur of Ryder's Hill, to the west-northwest. It rises fairly steeply from the Mardle Valley to the east, fairly steadily and then very gently from Mardle Head to the north, quite gently and then very gently from a col to the south-southeast and gently from the Western Wella Valley to the west. OS Maps shows its highest point to be on the top of a Bronze Age cairn (one of three on its summit area), at the OS 1:25000 map 495-metre spot height. However, it doesn't have a relative height of fifty feet and is linked to its parent peak by a very high col. A number of disused tin workings and several prehistoric remains are situated on its slopes; Buckfastleigh Moor is located on its southeastern flank and Snowdon Hole is sited on its east-southeastern one.

45. Hound Tor, 1,616.47 feet/492.70 metres, SX 628890, a somewhat elongated, partly rocky peak, which rises gently and then steadily from the Steeperton Valley to the west, gently from a col to the south-southwest, very gently from another to the east-northeast and variably from yet another to the north-northeast. OS Maps shows its rocky summit to be at the OS 1:25000 495-metre spot height. Gallaven Mire is situated at the foot of its south-southeastern flank, the prehistoric White Moor Stone on its northeastern one, White Moor Down Stone Circle at the foot of its north-northeastern one and Metheral Hill on its north-northwestern one.

46. [Lints Tor], 1,615.81 feet/492.50 metres, SX 579874, an ovalish-shaped, pretty rocky peak on a north-northwesterly-projecting spur of High Willhays (and, more immediately, Dinger Tor), well over a mile to the north. It rises steeply from the West Okement Valley to the west, quite gently from one of its tributary dales to the north and variably from another to the east. OS Maps shows its summit to be at the OS 1:25000 map 496-metre spot height, on the top of a spectacular conning tower-like rocky tor. However, it fails to be an independent hill by just 3.74 feet and is linked to its parent peak by a high col. A number of prehistoric hut circles are situated on its south-southwestern flank.

47. Shell Top, 1,615.49 feet/492.40 metres, SX 602645, an elongated, slightly rocky plateau, which rises steeply and then very gently from the Yealm Valley to the east, quite gently and then gently from Spanish Lake (via Lee Moor) to the west, quite gently and then very gently from the head of the Ford Valley to the south and very gently from Shavercombe Head to the north. The peak is better known for its south-southwestern spur of the same name, but OS Maps shows the hill's true summit to be at the OS 1:25000 map 493-metre rocky spot height, approximately 100 yards north of the 492-metre trig point. The hill is a significant watershed, with Shavercombe Head, Yealm Head, Broadall Head and the sources of Spanish Lake and the Hentor Brook all situated on its flanks. A series of prehistoric remains lie on its slopes, especially to the west and south, which include enclosed settlements, cairns, cists, stone hut circles and stone rows. Penn Moor is located on its south-southeastern flank, Yealm Steps waterfall at the foot of its east-southeastern one, Langcombe Hill on its east-northeastern one, Hentor Warren on its north-northwestern one, Hen Tor on its west-northwestern one and Penn Beacon on its south-southwestern one.

48=. [Black Dunghill], 1,611.22 feet/491.10 metres, SX 581774, a flattish-topped peak on a south-southwest-projecting spur of Cut Hill (and, more immediately, Conies Down and Maiden Hill), more than 3½ miles to the north-northeast. It rises very gently from the Blackbrook Valley to the south and gently from the same dale to the west and one of its tributary valleys to the east. OS Maps shows its summit to be just east-southeast of the OS 1:25000 map 492-metre spot height. However, it doesn't have a relative height of fifty feet and is linked to its parent peak by a high col. Blackbrook Head is situated at the foot of its north-northwestern flank.

48=. [Little Hound Tor], 1,611.22 feet/491.10 metres, SX 632899, an ovalish-shaped, partly rocky peak, which rises quite steeply from the Small Valley to the west, gently and then fairly steadily from Raybarrow Pool to the east and gently from a col to the south. OS Maps shows its summit to be approximately 50 yards north-northwest of the OS 1:25000 map 492-metre spot height. However, it fails to be an independent peak by just 1.77 feet and instead is an outlier of Cosdon Hill to the north-northeast, to which it's linked by a high col. White Moor Down Stone Circle is situated at the foot of its southern flank.

50. Five Barrows Hill, 1,610.89 feet/491.00 metres, SS 733368, a somewhat elongated peak, which rises steeply and then very gently from a tributary valley of the River Bray to the west, fairly steadily and then gently from the head of Span Bottom to the south, gently from Kinsford Water to the east and very gently from a tributary dale of the River Barle to the north. Although the peak is mistakenly named Five Burrows Hill on the OS 1:25000 map, nine Bronze Age barrows are actually situated upon it! OS Maps shows its summit to be approximately 100 yards east of the trig point, which is marked on the map as standing at 493 metres, although other evidence indicates that the highest spot is just north of the pillar. The county boundary with Somerset runs across its eastern and northern slopes, broadly from southeast to northwest. Yarde Down is located on its south-southwestern flank, Western Common on its west-southwestern one, the probably Bronze Age White Ladder double stone row on its northern one and Span Head at the foot of its southeastern one.

51. [Tom's Hill], 1,608.60 feet/490.30 metres, SX 642835, an elongated, flat-topped peak on a north-projecting spur of Sittaford Tor, to the west-southwest. It rises gently from a tributary dale of the North Teign River to the west and variably from the South Teign Valley to the east and a col to the north. OS Maps shows its summit to be approximately 25 yards south of the OS 1:25000 map 491-metre spot height. However, it fails to be an independent hill by only 7.67 feet and is linked to its parent peak by a high col. Fernworthy Forest covers most of its summit plateau and almost all of its eastern and northern slopes. Hemstone Rocks are situated on its east-southeastern flank; the Bronze Age Fernworthy Stone Circle, a series of stone hut circles and three stone rows are located on its east-northeastern one, at the foot of which is Fernworthy Reservoir, and Long Ridge is sited on its north-northwestern one.

52. [King Tor], 1,604.33 feet/489.00 metres, SX 708814, a small, partly rocky, flat-topped peak on a north-northeast-projecting spur of Hameldown Tor. It rises steeply and then quite gently from a headwater valley of the River Bovey to the east, quite steeply from another to the north and very gently from a col to the west-southwest. OS Maps shows its summit to be at the southern edge of the OS 1:25000 map 488-metre spot height. However, the hill doesn't have a relative height of fifty feet and is linked to its parent peak by a high col. King's Barrow, an impressive Bronze Age round cairn, is situated just to the east of its summit; Heathercombe Brake is located on its east-southeastern flank, with a wood on its lower slopes, and Hookney Down is sited on its western one.

53. [Water Hill], 1,602.69 feet/488.50 metres, SX 671812, a somewhat elongated, flat-topped peak, which rises variably from a depression to the west, the Walla Valley to the south and a headwater dale of the West Webburn River to the east. OS Maps shows its highest point to be on its summit Bronze Age round cairn, at the OS 1:25000 map 489-metre spot height. However, it fails to be an independent peak by just 0.79 feet and instead is an outlier of Assycombe Hill, to the northwest, to which it's linked by a high col. The remains of the old King's Oven tin-blowing house are situated on its eastern flank and a series of prehistoric stone hut circles are located on its western one.

54. Birch Tor, 1,600.39 feet/487.80 metres, SX 686815, an elongated, slightly rocky, flattish-topped peak, which rises steadily from the head of a tributary dale of the West Webburn River to the east, fairly steadily and then gently from another to the west and a col (via Headland Warren) to the south-southeast and quite steadily and then very gently from West Bovey Head to the north. OS Maps shows its summit to be both at the OS 1:25000 map 487-metre spot height and on the top of a prehistoric cairn approximately 50 yards to the north. The remains of Vitifer (tin) Mine are situated on its southwestern flank; Bennett's Cross wayside marker is located on its western one; Bush Down and the disused Bushdown Mine are sited on its west-northwestern one; several prehistoric hut circles lie on its north-northwestern one and East Bovey Head is found at the foot its northeastern one. The B 3212 runs across its western and northern slopes, broadly from southwest to northeast.

55. [Shapley Tor], 1,597.44 feet/486.90 metres, SX 699819, a wedge-shaped, slightly rocky, flattish-topped peak on a north-northeast-projecting spur of nearby Hookney Tor, to the south. It rises steadily from a headwater valley of the River Bovey to the east, fairly gently from a dry dale to the west and variably from the East Bovey Valley (via Shapley Common) to the north. OS Maps shows its summit to be just northeast of the OS 1:25000 map 487-metre spot height. However, the hill doesn't have a relative height of fifty feet and is linked to its parent peak by a high col. There are a series of prehistoric remains on its slopes; Coombe Down is situated on its east-northeastern flank and East Bovey Head is located at the foot of its west-northwestern one. The B 3212 cuts across its northern flank, broadly from west-southwest to east-northeast.

56. [Naker's Hill], 1,594.16 feet/485.90 metres, SX 643688, a small, irregular-shaped plateau, which rises fairly gently and then gently from the Avon Valley to the east, gently from Fishlake Mire to the south and very gently from the Blacklane Valley to the west. OS Maps shows its summit to be just northeast of the OS 1:25000 map 488-metre spot height. However, it fails by just 2.1 feet to be an independent peak and instead is an outlier of Ryder's Hill, to the east, to which it's linked by a high col. Swincombe Head is situated on its northwestern flank, Cater's Beam on its west-northwestern one and Blacklane Mire at the foot of its west-southwestern one.

57. [Standon Hill (Standon Down)], 1,589.24 feet/484.40 metres, SX 556815, a pretty rocky, round-topped peak at the end of a south-southwest-projecting spur of Cut Hill (and, more immediately, Walkham Head), approaching 2¾ miles to the east-northeast. It rises steeply and then gently from the Baggator Valley to the south, very gently from the same dale to the east and variably from the Tavy Valley to the north and west. OS Maps shows its summit to be at the OS 1:25000 map 485-metre spot height. However, it doesn't have a relative height of fifty feet and is linked to its parent peak by a very high col. A Bronze Age round cairn is situated at the southwestern end of its summit area and there are quite a number of prehistoric remains on its slopes.

58. [Skir Hill], 1,588.91 feet/484.30 metres, SX 645699, a wedge-shaped, flattish-topped peak, which rises very gently from the head of Deep Swincombe to the north and a col to the south-southwest and variably from the Swincombe Valley to the west. OS Maps shows its summit to be on the western edge of the OS 1:25000 map 487-metre spot height. However, it fails to be an independent hill by only 8.33 feet and instead is an outlier of Ryder's Hill (and, more immediately, Naker's Hill), just over a mile to the east-southeast, to which it's linked by a high col. Avon Head and Avon Head Mires are situated at the foot of its east-southeastern flank; the disused Hooten Wheals tin mine is located on its east-northeastern one; a Bronze Age ring cairn and cist and the remains of the Henroost tin mine are sited on its north-northeastern one; Skir Gut tinners' gully descends from its northern one and Ter Hill is found on its north-northwestern one.

59. [Ter Hill], 1,578.41 feet/481.10 metres, SX 642704, an elongated, flat-topped peak on a northwest-projecting spur of Ryder's Hill (and, more immediately, Skir Hill), approaching 1½ miles to the east-southeast. It rises very gently from Skir Gut to the east, fairly steadily and then gently from the Swincombe Valley to the west and variably from the latter to the north. OS Maps shows its summit to be on the western edge of the OS 1:25000 map 481-metre spot height. However, it doesn't have a relative height of fifty feet and is linked to its parent peak by very high col. Three waymarker crosses are situated on its flanks and two Bronze Age cists on its lower west-northwestern one.

60. Quickbeam Hill, 1,574.15 feet/479.80 metres, SX 652655, an elongated peak, which rises variably from the Avon Valley to the north, the Erme Valley to the west, a col to the south and Avon Dam Reservoir to the east. OS Maps shows its summit to be within the OS 1:25000 map small 480-metre contour ring, approximately 85 yards to the west-northwest of the 481-metre spot height. The latter is marked on top of the Bronze Age Western White Barrow, in the middle of which is the sixteenth century Petre's Cross. There are numerous Bronze Age remains on its flanks, including Eastern White Barrow, Knatta Barrow and Rider's Rings, an enclosed hut circle settlement. Brown Heath is situated on its west-northwestern flank.

61. Kennon Hill, 1,569.23 feet/478.30 metres, SX 642893, a wedge-shaped, slightly rocky, flat-topped peak, which rises fairly steadily and then very gently from the Blackaton Valley to the north, very gently from a col to the west and variably from the Gallaven Valley (via Rippator) to the south and Whitemoor Marsh to the east. OS Maps shows its summit to be at the OS 1:25000 map 478-metre spot height. An extensive Bronze Age stone hut circle settlement is situated on its southwestern flank, Raybarrow Pool at the foot of its north-northwestern one and Throwleigh Common on its northeastern one.

62. Butter Hill, 1,566.60 feet/477.50 metres, SS 711425, an elongated, twin-summited, flattish-topped peak, which rises steeply and then gently from Yarbury Combe to the south, steeply and then very gently from a tributary dale of the River Heddon to the west, very gently from a col to the east and variably from the Barbrook Valley to the north. OS Maps shows its highest point to be on its southeastern plateau, approximately 75 yards south-southeast of the OS 1:25000 map 478-metre spot height, although the latter marks its summit as 480 metres at SS 716425. The northwestern peak is also given a spot height of 480 metres at its trig point, although OS Maps computes it as 476.80 metres! The county boundary with Somerset runs across the eastern side of its southeastern summit area, broadly from south to north. There are numerous prehistoric remains on and around its summit areas, for example Longstone Barrow, Wood Barrow and Chapman Barrows. Barham Hill is situated on its northern flank; the source of the River Heddon is located on its northwestern one and the headwaters of the River Bray rise on its western one.

63. Belstone Tor, 1,561.35 feet/475.90 metres, SX 614919, an impressive, elongated, rocky peak, which rises steeply from the Taw Valley to the east and the East Okement Valley to the west and variably from a tributary dale of the latter (via Watchet Hill and Belstone Common) to the north and a col (via the Irishman's Wall) to the south-southwest. OS Maps shows its summit to be approximately 60 yards south-southwest of the OS 1:25000 map 479-metre spot height. There are several prehistoric remains on its slopes, including Nine Stones cairn circle on its northern flank; Higher Tor and Winter Tor are situated on its south-southwestern one; Scarey Tor is located on its west-northwestern one and the village of Belstone is sited on its north-northeastern one.

64. (Shoulsbarrow Hill), 1,559.38 feet/475.30 metres, SS 709392, an elongated, flat-topped peak, which rises fairly steeply and then very gently from Old Close Bottom to the north and Henthitchen Combe to the south, fairly steadily and then gently from Goat Combe to the west and variably from a headwater stream of Great Vintcombe to the east. OS Maps shows its summit to be approximately 150 yards east of the OS 1:25000 map 475-metre spot height. The hill is unnamed on the OS 1:25000 and 1:50000 maps, but it seems appropriate to name it after Shoulsbarrow Common (which occupies the upper northern part of it) and Shoulsbury (or Shoulsbarrow) Castle, an Iron Age hillfort (which is situated on the west-southwestern end of its summit plateau). A trig point is located at the western end of its summit area; Castle Common is sited to the south of its summit; Fullaford Down and Wallover Down are found on its west-southwestern flank; Roosthitchen lies on its east-northeastern one; Black Hill can be seen on its east-southeastern one and Henthitchen is positioned on its southeastern one. The county boundary with Somerset runs across its eastern flank, broadly from south to north.

65. [Green Hill], 1,553.81 feet/473.60 metres, SX 635678, a flattish-topped peak on a broad, south to south-southeast-curving spur of Ryder's Hill (and, more immediately, Skir Hill and Naker's Hill) approaching 1¾ miles to the west-northwest. It rises gently from the Blacklane Valley to the west, very gently from the head of Dry Lake to the south and variably from the Avon Valley to the east. OS Maps shows its summit to be approximately 75 yards northwest of the OS 1:25000 map 474-metre spot height. However, it doesn't have a relative height of fifty feet and is linked to its parent peak by a high col. A prehistoric cairn, which marks the northern end of the 2¼-mile-long Staldon (stone) Row, is situated on its summit area; the disused Red Lake China Clay Works is located on its southeastern flank and Fishlake Mire is sited at the foot of its east-northeastern one.

66. [Langcombe Hill], 1,550.20 feet/472.50 metres, SX 617656, a small, flat-topped, east-northeast-projecting spur of Shell Top, well over a mile to the west-southwest. It rises quite steeply and then very gently from the Erme Valley to the east, increasingly gently from Yealm Head to the south-southwest and very gently from Langcombe Head to the north and the head of a tributary stream of the Langcombe Brook to the west. OS Maps shows its summit to be approximately 100 yards southwest of the OS 1:25000 474-metre spot height. However, the land drops by only 7.55 feet to a very high col linking it to its parent peak. It is a significant watershed, with Langcombe Head, Yealm Head, Erme Head and the sources of the Bledge Brook and the Hortonsford Brook all situated on its slopes. Stall Moor is located on its south-southeastern flank and Stinger's Hill on its eastern one.

67. Rippon Tor, 1,548.23 feet/471.90 metres, SX 746755, a wedge-shaped, quite rocky peak, which rises increasingly steeply from the head of Ruddycleave Water to the west, fairly gently and then steeply from a col to the north-northeast and variably from the Ashburn Valley (via Halshanger Common) to the south and the Stig Valley (via Bagtor Down) to the east. OS Maps shows its summit to be on the top of its highest rocks, near the OS 1:25000 map trig point, which is marked as standing at 473 metres. A former logan stone, the Nutcracker (or Nutcrackers), is situated on its southwestern flank, several Bronze Age cairns lie on its summit area and a number of prehistoric settlements and stone hut circles are sited on its flanks. The source of the River Ashburn is found on its south-southeastern slopes and that of the River Stig at the foot of its east-northeastern ones. Mountsland Common is located on its south-southeastern flank, Horridge Common on its east-southeastern one and Buckland Common on its south-southwestern one. The B 3387 runs across its northern slopes, broadly from east to west.

68. [Crane Hill], 1,543.31 feet/470.40 metres, SX 621689, an irregular-shaped, small plateau, which rises gently and then very gently from the Erme Valley (via Great Gnats' Head) to the south, very gently from the head of the Blacklane Valley to the east and a col to the west and variably from the Swincombe Valley (via Foxtor Mires) to the north. OS Maps shows its summit to be at the OS 1:25000 471-metre spot height. However, it doesn't have a relative height of fifty feet and is an outlier of Ryder's Hill (and, more immediately, Skir Hill and Naker's Hill) to the east, to which it's linked by a high col. There are numerous Bronze Age remains on its southwestern slopes and disused tin workings on its northwestern ones. Erme Pits Hill is situated on its southern flank, Blacklane Mire on its southeastern one, Fox Tor on its north-northeastern one and Plym Head on its south-southwestern one.

69. [Great Gnats' Head], 1,542.98 feet/470.30 metres, SX 621679, a flattish-topped peak towards the southern end of the Crane Hill plateau, on a broad, south-projecting spur of Ryder's Hill (and, more immediately, Skir Hill, Naker's Hill and Crane Hill), nearly 2½ miles to the east-northeast. It rises quite gently and then very gently from the Plym Valley to the west and the Erme Valley to the south and very gently from the Blacklane Valley to the east. OS Maps shows its summit to be at the western side of the OS 1:25000 map 470-metre very small contour ring. However, it doesn't have a relative height of fifty feet and is linked to its parent peak by a very high col. There are quite a number of prehistoric remains on its slopes; Erme Pits Hill is situated on its south-southeastern flank; the Letter Box memorial is located on its eastern one and Calveslake Tor and Little Gnats' Head are sited on its west-southwestern one.

70. [Stinger's Hill], 1,528.22 feet/465.80 metres, SX 624656, a small, slightly rocky peak on an east-northeast-projecting spur of Langcombe Hill, to the west. It rises increasingly gently from the Erme Valley to the east, fairly gently from the head of the Bledge Valley to the south-southeast and quite gently from Hortonsford Bottom to the north. OS Maps shows its summit to be approximately 35 yards southwest of the OS 1:25000 map 468-metre spot height. However, the land drops by only 10.83 feet to a very high col linking it to its parent peak. There are a series of prehistoric remains on its south-southeastern and east-southeastern flanks, across which a stone row runs. Erme Plains and the remains of a blowing house are situated on its east-southeastern slopes.

71. White Tor (Cudliptown Down), 1,525.26 feet/464.90 metres, SX 542786, a roundish, fairly rocky peak, which rises quite steadily from the Colly Valley to the south and the Tavy Valley to the west, gently and then fairly steadily from the Youldon Valley to the north and very gently and then quite gently from a col to the east-northeast. OS Maps shows its summit to be just north of the OS 1:25000 map 468-metre spot height, within what is marked as a fort, but is now often believed to be a Neolithic enclosure. The hill is covered with prehistoric and historic remains, of which the Langstone, a tall Bronze Age menhir, at the foot of its eastern flank, is perhaps the most significant. There are one or two disused tin workings on its slopes; Langstone Moor is situated on its east-southeastern flank and Boulters Tor and Smeardon Down are located on its west-southwestern one.

72. [Higher Tor], 1,523.95 feet/464.50 metres, SX 612917, a small, pretty rocky peak on a south-southwest-projecting spur of Belstone Tor. It rises quite steeply from the Taw Valley to the east and the East Okement Valley to the west and increasingly steeply from a col to the south-southwest. OS Maps surprisingly shows its summit to be on the rocky outcrop symbol on the OS 1:25000 map and not within the tiny 470-metre contour ring just to the north-northeast. The fair discrepancy in the peak's altitude given by the online and paper versions of the OS's data might be explained by a limitation of the website's programme to calculate accurately the heights of rapidly-rising features. However, the land drops by only 9.84 feet to a very high col linking it to its parent peak. Winter Tor is situated on its west-southwestern flank.

73. [Pupers Hill], 1,520.01 feet/463.30 metres, SX 672674, a small, partly rocky peak, on a south-southeast-projecting spur of Ryder's Hill (and, more immediately, Snowdon), well over 1¼ miles to the north-northwest. It rises quite steeply and then gently from the Mardle Valley to the east-northeast and the head of the Dean Valley to the east-southeast, quite gently from the Western Wella Valley to the west and variably from Avon Dam Reservoir (via Hickaton Hill) to the south. Surprisingly, OS Maps shows its highest point to be approximately 65 yards southeast of the OS 1:25000 map 467-metre spot height on the Outer Pupers tor. However, the hill doesn't have a relative height of fifty feet and is linked to its parent peak by a high col. Pupers Rock is situated on its summit area and the Inner Pupers outcrop on its eastern flank. A Bronze Age cairn is also located on its summit area, there are others on its eastern flank and Buckfastleigh Moor is sited on its north-northeastern flank.

74. [Rowtor], 1,518.37 feet/462.80 metres, SX 593916, a very rocky peak on a north-northeast-projecting spur of High Willhays (and, more immediately, West Mill Tor), approaching 1¾ miles to the south-southwest. It rises quite steeply from the Moor Valley to the west, variably from the same dale to the north and fairly gently and then quite steeply from the East Okement Valley to the east. OS Maps shows its summit to be on the southern edge of the OS 1:25000 map 468-

metre spot height. The fair discrepancy in the peak's altitude given by the online and paper versions of the OS's data might be explained by a limitation of the website's programme to calculate accurately the heights of rapidly-rising features. The hill doesn't have a relative height of fifty feet and is linked to its parent peak by a high col.

75. (Ugborough Hill), 1,513.78 feet/461.40 metres, SX 652626, a slightly rocky, wedge-shaped peak, which rises gently and then pretty steadily from the Red Valley to the east, very gently from a col to the north and variably from the Erme Valley (via Harford Moor) to the west and a dry dale (via Piles Hill) to the south. OS Maps shows its highest point to be on its summit cairn, just west of the OS 1:25000 map 464-metre spot height and a little north-northwest of the trig point, which is also marked as standing at 464 metres. The hill is unnamed on the OS 1:25000 and 1:50000 maps, but it seems appropriate to name it after Ugborough Moor, which covers most of its slopes. The Bronze Age Three Barrows are situated on its summit area (the highest of which marks the hill top) and other prehistoric remains abound on its slopes. Glasscombe Ball is located on its south-southeastern flank; Glaze Head is sited on its southeastern one; Hickley Plain and Brent Fore Hill lie on its east-southeastern one; Leftlake Mires are found on its north-northwestern one and Sharp Tor can be seen on its south-southwestern one.

76. [Stannon Tor], 1,513.45 feet/461.30 metres, SX 646811, a rocky south-southeast-projecting spur of White Ridge, to the north-northeast. It rises very gently and then fairly steadily from the East Dart Valley to the south and variably from the same dale to the west and the Stannon Valley to the east. OS Maps shows its summit to be towards the southern end of the OS 1:25000 map 460-metre contour ring, but the land drops by only 10.83 feet to a very high col linking it to its parent peak. Hartland Tor and several prehistoric hut circles are situated on its south-southwestern flank.

77. Challacombe Down, 1,510.17 feet/460.30 metres, SX 689803, a somewhat elongated, partly rocky peak, which rises steeply from a tributary valley of the West Webburn River to the east, fairly steadily and then quite steeply from the West Webburn Valley to the west, fairly steadily and then gently from the latter to the south and quite gently from a col to the north. OS Maps shows its summit to be towards the southern end of the OS 1:25000 map 460-metre contour ring. A prehistoric triple stone row is situated on its northern flank; stone hut circles are located on its west-northwestern one; the disused Golden Dagger Tine Mine and medieval field boundaries are sited on its south-southwestern one and the remains of the medieval village of Challacombe are found at the foot of its south-southeastern one.

78. [Oke Tor], 1,507.22 feet/459.40 metres, SX 612900, an elongated, rocky peak on a north-northwest-projecting spur of Black Hill (and, more immediately, Okement Hill), approaching 3½ miles to the south. It rises fairly steadily from the East Okement Valley to the west, gently from a col to the north and variably from the Taw Valley to the east. OS Maps shows its summit to be approximately 35 yards northwest of the OS 1:25000 map 466-metre spot height. The considerable discrepancy in the peak's altitude given by the online and paper versions of the OS's data might be explained by a limitation of the website's programme to calculate accurately the heights of rapidly-rising features. It doesn't have a relative height of fifty feet and is linked to its parent peak by a high col.

79. [Metheral Hill], 1,504.92 feet/458.70 metres, SX 624896, a short ridge on a north-northwest-projecting spur of Hound Tor, to the southeast. It rises quite steeply from the Steeperton Valley to the west, quite steeply and then very gently from Taw Marsh to the north-northwest and gently from the Small Valley to the east. OS Maps shows its summit to be at the OS 1:25000 459-metre spot height, but the land drops by only 11.81 feet to a very high col linking it to its parent peak.

80. Chinkwell Tor, 1,496.06 feet/456.00 metres, SX 729782, a somewhat elongated, rocky tor, which rises steadily and then quite steeply from the East Webburn Valley to the west, variably from one of its tributary dales to the south and gently and then fairly steadily from cols to the east and north. OS Maps shows its highest point to be on its summit tor, towards the northeastern end of the OS 1:25000 map 450-metre contour ring. There are a number of Bronze Age stone hut circles on its slopes; Bell Tor, a series of rocky outcrops, is situated on its south-southeastern flank; Slades Well is located on its northern-northwestern one and there is a fair amount of woodland on its western slopes.

81. Higher Hartor Tor, 1,488.19 feet/453.60 metres, SX 599686, an elongated, partly rocky, flat-topped peak, which rises fairly steadily from the Narrator Valley to the west, gently and then very gently from a col to the north-northeast, very gently from a depression to the east and variably from the Plym Valley to the south. OS Maps shows its summit to be approximately 50 yards north-northeast of the OS 1:25000 map 454-metre spot height. Its main tor is situated on its upper southern flank and Lower Hartor Tor is located on its south-southeastern one. Two Bronze Age cairns (one of them being Eylesbarrow) are sited on its summit area; the disused Eylesbarrow Tin Mine is found on its upper south-southwestern flank and a whole series of prehistoric remains are situated on its slopes, especially to the south-southwest.

82. [Brat Tor], 1,486.55 feet/453.10 metres, SX 539856, a partly rocky peak at the end of a west-southwest-projecting spur of Great Links Tor, just over a mile to the northeast. It rises fairly steadily and then steeply from the Lyd Valley to the west, variably from one of its tributary dales so the north and fairly steadily and then quite steeply from the Doetor Valley to the south. OS Maps shows its summit to be on the northeastern end of its main rocky outcrop, approximately 35 yards northeast of the OS 1:25000 map 452-metre spot height. However, the land drops by only 11.48 feet to a very high col linking it to its parent peak. The impressive granite Widgery Cross is situated on its summit area.

83. South Hessary Tor, 1,474.08 feet/449.30 metres, SX 594730, an elongated, twin-summited, small plateau, which rises fairly steadily and then very gently from the Blackbrook Valley to the north, increasingly gently from the Meavy Valley to the west, very gently from a dry dale to the south and variably from a tributary valley of the Blackbrook River to the east. OS Maps shows its main summit to be on its north-northwestern top, approximately 65 yards south-southeast of the OS 1:25000 map 451-metre spot height. The highest point of its subsidiary (south-southeastern) top is computed to be 1,472.44 feet, on its summit tor, inside the OS 1:25000 map 450-metre tiny contour ring, at grid reference SX 597723. There are areas of woodland on some of its lower slopes; the remains of a large prehistoric settlement are situated on its west-southwestern flank; Devil's Bridge is located on its western one; Soldiers' Pond is sited on its west-northwestern one and the village of Princetown is found at the foot of its northwestern one. The B 3212 runs across its western and northern flanks, broadly from southwest to northeast.

84. Merripit Hill, 1,473.10 feet/449.00 metres, SX 657803, an elongated, flattish-topped peak, which rises fairly steadily and then quite gently from the Stannon Valley to the west, very gently from a col to the north-northeast, fairly steadily and then gently from the head of a tributary dale of the East Dart River to the south and variably from another such tributary valley to the east. OS Maps shows its summit to be at the OS 1:25000 map 449-metre spot height. The remains of Second World War military buildings are situated on its summit area; there are several disused quarries on its slopes and a prehistoric hut circle is located on its east-southeastern flank. The B 3212 runs across its southern and eastern slopes, broadly from southwest to northeast.

85=. Haytor Rocks, 1,471.46 feet/448.50 metres, SX 757770, a wedge-shaped, fairly rocky, twin-summited peak, which rises gently from a col to the north-northeast and variably from the Sig Valley to the south, the Lemon Valley to the east and the Becka Valley to the west. OS Maps shows its highest point to be on the top of its bulky west-southwestern summit tor, at the OS 1:25000 map 457-metre spot height. Its subsidiary top is an even more impressive tor and is approximately 150 yards to the east-northeast, computed by OS Maps to be 1,432.74 feet high. The large discrepancy in the peak's altitude given by the online and paper versions of the OS's data might be explained by a limitation of the website's programme to calculate accurately the heights of rapidly-rising features. Bag Tor and disused mines are situated on its south-southeastern flank; the source of the River Lemon is located at the foot of its east-northeastern one and the disused Haytor Quarries are sited on its north-northeastern one. The B 3387 runs across its southern flank and down its eastern one, essentially from west-southwest to east-northeast.

85=. Roos Tor, 1,471.46 feet/448.50 metres, SX 543766, an elongated, fairly rocky peak, which rises fairly steadily from a dry dale to the west, gently from a col to the south-southwest, very gently from another to the north-northeast and variably from the Walkham Valley to the east. OS Maps shows its highest point to be on the top of its summit tor, at the OS 1:25000 map 454-metre spot height. The fair discrepancy in the peak's altitude given by the online and paper versions of the OS's data might be explained by a limitation of the website's programme to calculate accurately the heights of rapidly-rising features. A logan stone lies on its summit and a number of Bronze Age settlements and stone hut circles are situated on its flanks.

87. [Raddick Hill], 1,464.24 feet/446.30 metres, SX 596715, an elongated, partly rocky, twin-summited peak, which rises increasingly gently from Newleycombe Lake to the south and variably from the Meavy Valley (via Cramber Tor) to the west and the Strane Valley to the east. OS Maps shows its main (northeastern) summit to be approximately 50 yards south-southwest of a pool in the northwestern section of the OS 1:25000 map 440-metre elongated contour ring. Its subsidiary (southwestern) top is computed to be 4.59 feet lower and positioned about 100 yards east of the trig point, which is marked as standing at 445 metres. However, it fails to be an independent hill by only 9.64 feet and instead is an outlier of South Hessary Tor, approaching a mile to the north, to which it's linked by a high col. There are quite a number of prehistoric remains and several disused tin workings on its slopes and its lower west-southwestern ones are largely wooded.

88. Great Staple Tor, 1,463.25 feet/446.00 metres, SX 542760, a somewhat elongated, rocky, roundish-topped peak, which rises quite gently from a col to the west-northwest, gently from another to the north-northeast and variably from the Walkham Valley to the east and one of its tributary dales (via Middle and Little Staple Tor) to the south-southwest. OS Maps shows its highest point to be on the top of the unusual tall rock stack on its summit tor, at the OS 1:25000 map 455-metre spot height. The large discrepancy in the peak's altitude given by the online and paper versions of the OS's data might be explained by a limitation of the website's programme to calculate accurately the heights of rapidly-rising features. Merrivale and Merrivale Quarry are situated on its south-southeastern flank; there are one or two disused tin workings on its slopes and the B 3357 runs across its southern flank, broadly from east to west.

89. [Ger Tor], 1,458.01 feet/444.40 metres, SX 546834, a small, pretty rocky peak on a south-projecting spur of Rattlebrook Hill (and, more immediately, Sharp Tor and Hare Tor), well over 1¼ miles to the north-northeast. It rises fairly steadily from the Willsworthy Valley to the west, gently from a tributary dale of the River Tavy to the east and variably from the Tavy Valley (via its own main tor) to the south. OS Maps shows its summit to be just east of the OS 1:25000 map 445-metre spot height. However, the land drops by only 6.24 feet to a very high col linking it to its parent peak. There are numerous prehistoric settlements on its slopes; Nat Tor is situated on its southern flank and Nattor Down is located on its south-southwestern one.

90. Bellever Tor, 1,443.90 feet/440.10 metres, SX 644764, an oval-shaped, very rocky, roundish-topped peak, which rises fairly gently and then quite steeply from a tributary valley of the East Dart River to the east, fairly gently and then steadily from the Cherry Valley to the west, very gently and then quite gently from a col to the north-northeast and variably from the West Dart Valley to the south. OS Maps shows its highest point to be on its summit tor by the OS 1:25000 map trig point, which is marked as standing at 443 metres. Most of its western and eastern flanks are wooded; the prehistoric Dunnabridge Pound is situated on its southern one and its slopes are covered by Bronze Age remains, including stone hut circles, a cairn circle, a settlement, cairns and cists. The B 3357 runs across its lower southern flank, broadly from east-southeast to west-northwest.

91. Cox Tor, 1,442.26 feet/439.60 metres, SX 530761, an ovalish-shaped, fairly rocky peak, which rises fairly gently from a col to the east, gently and then fairly steadily from another to the south and variably from the Colly Valley to the north and the Tavy Valley to the west. OS Maps shows its summit to be at the OS 1:25000 map trig point, which is marked as standing at 442 metres. There are numerous prehistoric remains, including cairns, hut circles and settlements, on its slopes; Great Combe Tor is situated on its north-northwestern flank and the village of Peter Tavy is located at the foot of its northwestern one. The B 3357 runs across its southern slopes, broadly from west-southwest to east-northeast.

92. Honeybag Tor, 1,440.94 feet/439.20 metres, SX 728786, an elongated, pretty rocky peak, which rises fairly steadily and then quite steeply from the East Webburn Valley to the west, steadily from one of its tributary dales to the east, quite gently from another to the north and fairly gently from a col to the south-southeast. OS Maps shows its highest point to be on the top of its summit tor, approximately 35 yards south of the OS 1:25000 map 445-metre spot height. The fair discrepancy in the peak's altitude given by the online and paper versions of the OS's data might be explained by a limitation of the website's programme to calculate accurately the heights of rapidly-rising features. Several Bronze Age stone hut circles and a fair amount of woodland are situated on its slopes.

93. [Sourton Tors], 1,437.66 feet/438.20 metres, SX 543898, a peak of rocky outcrops on a west-northwest-projecting spur of Amicombe Hill (and more immediately Corn Ridge), well over two miles to the southeast. It rises increasingly steeply from the Lew Valley to the north (via Prewley Moor) and west, increasingly gently from a tributary dale of the Crandford Brook to the south and very gently from a dry valley to the east. OS Maps shows its summit to be at the OS 1:25000 map trig point, which is marked as standing at 440 metres. However, it fails to be an independent hill by just 3.74 feet and is linked to its parent peak by a high col. The disused Sourton Quarry and the village of Bridestowe are situated on its west-southwestern flank; the village of Sourton is located on its west-northwestern one; the source of the River Lew is sited on its north-northeastern one and Meldon Reservoir is found at the foot of its east-northeastern one. The A 386 runs across its lower western slopes, broadly from south-southwest to north-northeast.

94. [Fox Tor], 1,432.41 feet/436.60 metres, SX 625697, a small, partly rocky peak on a northeast-projecting spur of Ryder's Hill (and, more immediately, Skir Hill, Naker's Hill and Crane Hill), approaching 2¼ miles to the east-southeast. It rises gently from Fox Tor Girt tinners' gully to the east, very gently and then fairly steeply from the Swincombe Valley to the north and very gently from a tributary dale of the latter to the west. OS Maps shows its rocky summit to be approximately

50 yards southwest of the OS 1:25000 map 438-metre spot height. However, the land drops by only 5.24 feet to a very high col linking it to its parent peak. The remains of Childe's (a traveller's) Tomb are situated on its northern flank.

95. Great Nodden, 1,431.10 feet/436.20 metres, SX 539874, an elongated peak, which rises steeply from the Lyd Valley to the east, steeply and then quite steeply from the same dale to the south, fairly steeply from the Crandford Valley to the west and very gently from a tributary dale of the latter to the north-northeast. OS Maps shows its summit to be at the northern edge of the OS 1:25000 map 437-metre spot height. A prehistoric cairn is situated on its summit area, the village of Vale Down on its southwestern flank and Southerly Down on its west-northwestern one. The King Wall runs across its western and northern slopes, broadly from south-southwest to north-northeast, and the A 386 crosses its lower northwestern ones, essentially from south to north.

96. Black Down, 1,430.12 feet/435.90 metres, SX 581922, an oval-shaped peak, which rises quite steeply from the West Okement Valley to the west (via the huge Meldon Quarry) and north and the Moor Valley to the east and variably from a col to the south. OS Maps shows its summit to be at the OS 1:25000 map 436-metre spot height. There are disused quarries on its east-northeastern flank; Okehampton (army) Camp is situated on its north-northeastern one; Okehampton is located at its foot and the site of the Saxongate medieval settlement is found on its northern one. The A 30 and the Dartmoor Railway run across its lower (fairly wooded) northeastern and northern slopes, broadly from west-southwest to east-northeast.

97. Easdon Tor (Easdon Down), 1,428.81 feet/435.50 metres, SX 729823, an isolated, elongated, partly rocky, roundish-topped peak, which rises fairly steadily and then quite steeply from a tributary valley of the River Bovey to the west and variably from the same dale to the north, another tributary to the east and the Hayne Valley to the south. OS Maps shows its summit to be on the top of its tor, at the western end of a 600-yard-long, east to west-running ridge, at the OS 1:25000 map trig point, which is marked as standing at 439 metres. The impressive Whooping Rock is situated just below the pillar; Easdon Hill is located at the eastern end of its greater summit area; a series of Bronze Age stone hut circles are sited on its flanks and there is a fair amount of woodland on its lower slopes.

98. North Molton Ridge, 1,426.18 feet/434.70 metres, SS 778325, an elongated, pretty flat-topped peak, which rises fairly steeply and then gently from the Mole Valley to the north, increasingly gently from one of its tributary dales to the west and one of the River Yeo to the south and fairly gently and then very gently from Litton Water to the east. OS Maps shows its summit to be approximately 75 yards northwest of the OS 1:25000 map trig point, which is marked as standing at 435 metres. Two prehistoric bowl barrows are situated on its summit area and there are small patches of woodland on some of its lower slopes. Barcombe Down is located on its west-southwestern flank and Twitchen Ridge on its eastern one. The source of the River Mole can be found at the foot of its north-northeastern flank.

99. [Harter Hill], 1,425.20 feet/434.40 metres, SX 601911, an elongated, partly rocky, roundish-topped peak on a north-projecting spur of East Mill Tor, to the south. It rises fairly gently from the East Okement Valley to the east and variably from the Black-a-ven Valley to the west and north. OS Maps shows its summit to be at the OS 1:25000 map 436-metre spot height. However, the hill doesn't have a relative height of fifty feet and is linked to its parent peak by a very high col. There are two small woods on its east-southeastern slopes.

100. Shovel Down, 1,417.65 feet/432.10 metres, SX 658857, a somewhat elongated, flattish-topped peak, which rises very gently from a col to the east-northeast and a tributary dale of the North Teign River to the west and variably from the North Teign Valley to the north and a tributary stream of Fernworthy Reservoir to the south. OS Maps shows its summit to be just south of the middle of the OS 1:25000 map 430-metre contour ring. The hill is covered with prehistoric remains, including the Fourfold Circle, the Long Stone, Three Boys standing stone and a series of stone rows. Teign-e-ver Clapper Bridge is situated at the foot of its north-northwestern flank.

101. Corndon Tor (Corndon Down), 1,410.10 feet/429.80 metres, SX 685741, a south to north-northwest-running, partly rocky ridge, which rises quite gently from the head of Simon's Lake to the south and a col to the west-southwest and variably from another to the north and the West Webburn Valley to the east. OS Maps shows its highest point to be on the top of its summit rocks, at the OS 1:25000 map 434-metre spot height. The fair discrepancy in the peak's altitude given by the online and paper versions of the OS's data might be explained by a limitation of the website's programme to calculate accurately the heights of rapidly-rising features. Several Bronze Age cairns are situated on its summit area and flanks and one or two stone hut circles are located on its lower south-southeastern slopes.

102. [Hoccombe Hill], 1,408.79 feet/429.40 metres, SS 764438, an elongated, flattish-topped peak, which rises increasingly gently from the Farley Valley to the west, gently and then very gently from the head of a tributary dale of Hoccombe Combe to the east and very gently from a col to the south and the head of Lank Combe (via Brendon Common) to the north. OS Maps shows its summit to be just southeast of the OS 1:25000 map 429-metre spot height. However, it doesn't have a relative height of fifty feet and is an outlier of Exe Plain (and, more immediately, Hoar Tor), approaching 1¼ miles to the south-southwest, to which it's linked by a high col. There are numerous prehistoric remains on its slopes; Clannon Ball is situated on its west-southwestern flank; Pig Hill is located on its northwestern one; Shilstone Hill is sited on its north-northwestern one and Withycombe Ridge is found on its north-northeastern one. The B 3223 cuts across its western flank, broadly from south-southeast to north-northwest.

103. Top Tor, 1,406.50 feet/428.70 metres, SX 736762, an elongated, rocky peak, which rises gently and then very gently from a col to the south-southwest, very gently from the head of a tributary dale of the East Webburn River to the north-northwest and variably from the Becka Valley to the east and the East

Webburn Valley (via Hollow Tor) to the west. OS Maps shows its highest point to be on the top of its summit tor, at the OS 1:25000 map 432-metre spot height. There are a number of prehistoric remains on its slopes, including Foale's Arrishes, a Bronze Age settlement, on its south-southeastern ones. Pil Tor and Blackslade Down are situated on its south-southwestern flank; Tunhill Rocks are located on its southwestern one; the Shovel Stone and Rugglestone Rock are sited on its west-northwestern one and Blackslade Mire lies at the foot of its southern one.

104. [Penn Beacon], 1,403.22 feet/427.70 metres, SX 599629, a small, fairly rocky peak on a south-projecting spur of Shell Top. It rises steadily from Cholwichtown China Clay Works to the west, fairly steadily and then quite gently from the infant Ford Valley to the east and variably from the village of Cornwood to the south-southeast. OS Maps shows its summit to be at the OS 1:25000 map 427-metre spot height at the trig point, but the map itself and other evidence indicates that a prehistoric cairn, just to the west-southwest, is its highest point. However, the hill doesn't have a relative height of fifty feet and is linked to its parent peak by a very high col. There a series of prehistoric remains on its slopes, including enclosed settlements, cairns and stone rows; Rook Tor is situated on its south-southwestern flank and its lower south-southwestern slopes are heavily wooded.

105. [Kestor Rock], 1,401.57 feet/427.20 metres, SX 665862, an impressive tor topping an east-northeast to north-curving spur of Shovel Down, to the west-southwest. It rises steeply and then gently from the North Teign Valley to the north, fairly steeply and then quite gently from the South Teign Valley to the east and very gently from a col to the south-southwest. OS Maps shows its summit to be on the top of the rocky outcrop at the OS 1:25000 map 437-metre spot height. The large discrepancy in the peak's altitude given by the online and paper versions of the OS's data might be explained by a limitation of the website's programme to calculate accurately the heights of rapidly-rising features. It fails by only 6.37 feet to be an independent hill and is linked to its parent peak by a high col. Chagford Common covers most of its upper slopes; a number of Bronze Age hut circles, including Round Pound, are situated on its flanks; Middle Tor and Frenchbeer Rock are located on its south-southeastern one and there is a fair amount of woodland on some of its lower slopes.

106. [Down Ridge], 1,399.93 feet/426.70 metres, SX 653716, a flat-topped peak on a broad, east-projecting spur of Ryder's Hill (and, more immediately, of Skir Hill and Ter Hill), well over 1½ miles to the south-southeast. It rises quite steeply and then very gently from the O Valley to the east, very gently from the same dale to the south and variably from the West Dart Valley to the north. OS Maps shows its summit to be approximately 35 yards west-southwest of the OS 1:25000 map 427-metre spot height. However, it doesn't have a relative height of fifty feet and is connected to its parent peak by a very high col. There are numerous prehistoric remains, two stone crosses, two blowing houses and two disused mines on its slopes and the hamlet of Hexworthy is situated on its northern flank.

107. [Crockern Tor], 1,392.39 feet/424.40 metres, SX 616762, a small, flattish-topped peak on a south-projecting spur of Longaford Tor (and, more immediately, Littaford Tors), more than a mile to the north. It rises increasingly gently from the West Dart Valley to the west, and variably from the Muddilake Valley to the south and the Cherry Valley to the east. OS Maps shows its summit to be at the OS 1:25000 map 425-metre spot height, but the land drops by only 8.20 feet to a very high col linking it to its parent peak. Its main tor is situated on its south-southwestern flank and is the site of the former Stannary Parliament. A number of pillow mounds are located on its west-southwestern flank; Muddilake and several prehistoric hut circles lie on its southeastern one and the B 3212 runs across its lower southern and eastern slopes, broadly from west-southwest to east-northeast.

108. [Easdon Hill], 1,391.40 feet/424.10 metres, SX 733823, the eastern end of a 600-yard ridge at the top of Easdon Down. It rises fairly steadily from the Hayne Valley to the south, fairly gently from the head of a tributary dale of the River Bovey to the north and variably from another such valley to the east. OS Maps shows its summit to be at the eastern edge of the OS 1:25000 map 425-metre spot height, but the land drops by only 8.53 feet to a very high col linking it to Easdon Tor, at the western end of the ridge, of which it's an outlier. Figgie Daniel, an impressive tor, is situated on its east-southeastern flank and there is a considerable amount of woodland on its lower eastern and north-northeastern slopes.

109. Saddle Tor, 1,390.09 feet/423.70 metres, SX 750764, a small, roundish, rocky peak, which rises gently and then quite steeply from the Becka Valley to the west and the head of a tributary dale of the River Sig to the east, very gently and then quite steeply from a col to the east-northeast and variably from another one to the southwest and the head of the Sig Valley to the south. OS Maps shows its highest point to be on the top of its summit tor, at the OS 1:25000 map 428-metre spot height. The fair discrepancy in the peak's altitude given by the online and paper versions of the OS's data might be explained by a limitation of the website's programme to calculate accurately the heights of rapidly-rising features. Two prehistoric hut circles are situated near the foot of its eastern flank and the B 3387 runs across its southern and eastern ones, broadly from west-southwest to east-northeast.

110. Emsworthy Rocks, 1,384.51 feet/422.00 metres, SX 752768, a small, roundish, pretty rocky peak, which rises gently and then fairly steadily from a col to the east-northeast, very gently from another to the south, very gently and then fairly gently from a depression to the north and variably from the Becka Valley to the west. Although the hill is unnamed on the OS 1:25000 and 1:50000 maps, the tor and the innumerable rocks surrounding it are generally known as Emsworthy Rocks. OS Maps shows its highest point to be on the top of its summit tor, at the OS 1:25000 map 420-metre spot height. Disused quarries are sited on its slopes and several Bronze Age stone hut circles on its west-northwestern ones. The B 3387 runs across its southeastern ones, broadly from south-southwest to north-northeast.

111=. [Chittaford Down], 1,382.22 feet/421.30 metres, SX 632788, a flat-topped peak on a wide, south-projecting spur of Broad Down, more than 1½ miles to the west-northwest. It rises very gently from a tributary dale of Braddon Lake to the north and variably from the Gawler Valley to the east and the Cherry Valley (via Arch Tor) to the south. OS Maps shows its summit to be approximately in the middle of the OS 1:25000 map 420-metre contour ring. However, it doesn't have a relative height of fifty feet and is linked to its parent peak by a very high col. There are a considerable number of prehistoric sites on its flanks, including the Roundy Park enclosure and settlement remains to the northeast.

111=. Lakehead Hill, 1,382.22 feet/421.30 metres, SX 643775, a somewhat elongated, flat-topped peak, which rises gently from a tributary dale of the Cherry Brook to the south, very gently from a col to the southeast and variably from another to the west, the East Dart Valley to the east and the Gawler Valley to the north. OS Maps shows its summit to be approximately in the middle of the southern bulge of the OS 1:25000 map 420-metre contour ring, in a large clearing in an extensive forestry plantation. There are numerous Bronze Age remains situated on its summit plateau and flanks, including Kraps Ring settlement, a stone row, cairn circles, cists and a stone hut circle. There are several disused quarries on its slopes; the B 3212 runs across its western and northern ones, broadly from south-southwest to north-northeast, and the hamlet of Postbridge is located at the foot of its north-northeastern flank.

113=. [Soussons Down], 1,381.89 feet/421.20 metres, SX 676803, a somewhat elongated peak, which rises fairly steeply and then gently from the Walla Valley to the west, fairly steadily from a tributary dale of the West Webburn River to the east and variably from a col to the south. OS Maps shows its summit to be just south-southeast of the OS 1:25000 map 420-metre spot height. However, it fails by just 6.69 feet to be an independent hill and instead is an outlier of Assycombe Hill (and, more immediately, Water Hill), well over 1¼ miles to the north-northwest, to which it's linked by a high col. Although most of the hill is heavily wooded, its summit and the majority of its northern hemisphere are on open ground. Four Bronze Age round barrows (the Red Barrows) and the Ringastan cairn circle are situated on its southern flank and Soussons Warren is located on its south-southeastern one.

113=. Thornworthy Tor (Thornworthy Down), 1,381.89 feet/421.20 metres, SX 664851, a small, roundish, very rocky peak, which rises fairly steadily from a tributary valley of Fernworthy Reservoir to the west, quite gently from another to the south, very gently and then gently from a col to the north and variably from a tributary dale of the South Teign River to the east. OS Maps shows its summit to be at the OS 1:25000 map 424-metre spot height. A large, remarkably-perched logan stone is situated on the top of the specific tor itself and several prehistoric settlements are located on its northeastern flank.

115. [Rippator (Rival Tor)], 1,379.92 feet/420.60 metres, SX 642881, a small, somewhat elongated, partly rocky peak, with a low rocky tor, on a south-projecting spur of Kennon Hill, to the north. It rises quite gently from the Gallaven Valley to the south, very gently and then gently from the same dale to the west and very gently and then quite gently from a dry dale to the east. OS Maps shows its summit to be at the OS 1:25000 map 421-metre spot height, but the land drops by only 11.81 feet to a very high col linking it to its parent peak.

116. Laughter Tor, 1,378.94 feet/420.30 metres, SX 653757, an ovalish-shaped, rocky peak, which rises increasingly steeply from a col to the west-northwest, gently from a tributary dale of the East Dart River to the north-northeast and variably from the West Dart Valley to the south and the East Dart Valley to the east. OS Maps shows its highest point to be on the top of its summit tor, at the OS 1:25000 map 420-metre spot height. Most of its northern and eastern and lower southeastern flanks are heavily wooded. There are numerous Bronze Age remains on its (mainly southern) slopes, including Dunnabridge Pound; Huccaby Ring enclosure; Outer Huccaby Ring enclosed settlement; the tall menhir, the Loughtor Man; a double stone row and a series of stone hut circles. Huccaby Tor is situated on its south-southeastern flank.

117. [Sharp Tor], 1,375.98 feet/419.40 metres, SX 650618, a small, partly rocky, flat-topped peak on a south-southwest-projecting spur of Ugborough Hill. It rises steeply from the Erme Valley to the west, fairly steadily and then gently from the head of the Piles Valley to the south and very gently from Glaze Head to the east. OS Maps shows its highest point to be approximately 75 yards north-northwest of the prehistoric cairn on its summit area, but the land drops by only 12.13 feet to a very high col linking it to its parent peak. The rocky tor itself is about 225 yards southwest of its summit; Ugborough Moor covers most of its upper slopes and Piles Copse, a rare high-level ancient woodland remnant, is situated on its lower west-northwestern flank.

118. [Doe Tor], 1,372.70 feet/418.40 metres, SX 541848, a small, rocky peak on a west to west-northwest-curving spur of Rattlebrook Hill (and, more immediately, Sharp Tor), approaching a mile to the east-northeast. It rises increasingly steeply from the Lyd Valley to the west, fairly steadily from the Walla Valley to the south and very gently from the Doetor Valley to the north. OS Maps shows its highest point to be at the western edge of the OS 1:25000 map 425-metre spot height, which is on the top of its summit tor. The considerable discrepancy in the peak's altitude given by the online and paper versions of the OS's data might be explained by a limitation of the website's programme to calculate accurately the heights of rapidly-rising features. The hill doesn't have a relative height of fifty feet and is linked to its parent peak by a high col. Doetor Common is situated on its west-southwestern flank; a Bronze Age round cairn and cist are located on its lower eastern one and Wallabrook Head is sited at the foot of its south-southeastern one.

119. [Hickaton Hill], 1,362.86 feet/415.40 metres, SX 672662, a small, slightly rocky, flattish-topped peak on an east-southeast-projecting spur of Ryder's Hill (and, more immediately, Snowdon and Pupers Hill), approaching two miles to the north-northwest. It rises increasingly gently from the head of the Brockhill Valley (via Brockhill Mire) to the east, fairly steeply from the Avon Dam Reservoir to the south and variably from the Western Wella Valley to the west. OS Maps shows its summit to be approximately 30 yards southeast of the OS 1:25000 map 417-metre spot height, but the land drops by only 6.23 feet to a very high col linking it to its parent peak. There are a fair number of prehistoric remains on its flanks, including Bronze Age pounds and stone hut circles.

120=. Buttern Hill, 1,358.60 feet/414.10 metres, SX 652886, a somewhat elongated, partly rocky, flat-topped peak, which rises gently from a col to the west, very gently from the Walla Valley (via Gidleigh Common) to the south and the Forder Valley to the north and variably from the Moortown Valley to the east. OS Maps shows its summit to be approximately 65 yards south of the OS 1:25000 map 413-metre spot height. There are a series of prehistoric remains on its slopes and Whitemoor Marsh is situated at the foot of its north-northwestern flank.

120=. Royal Hill, 1,358.60 feet/414.10 metres, SX 611727, a small, elongated, slightly rocky, twin-summited plateau, which rises very gently from the Strane Valley to the south and Devonport Leat to the west and variably from the West Dart Valley to the north and the head of Rue Lake to the east. OS Maps shows its highest point to be on its southeastern top, a little to the southeast of the middle of the OS 1:25000 map 410-metre contour ring. Its subsidiary (northwestern) summit is shown by OS Maps to be approaching half a mile away, on a small tor to the southwest of the OS 1:25000 map 412-metre spot height, with an altitude of 1,350.39 feet. There are a series of Bronze Age cairns and cists on its flanks, including the Crock of Gold cairn circle and cist on its eastern side. Strane Head is situated on its southern flank, Cholake Head on its eastern one and Swincombe Intake Works at the foot of its east-southeastern one.

122. Black Hill, 1,357.61 feet/413.80 metres, SX 760786, an elongated, partly rocky peak, which rises fairly steeply and then gently from the head of Reddaford Water to the east, fairly steeply from the Becka Valley to the west, variably from the same dale to the north and very gently from a col to the south-southwest. OS Maps shows its summit to be at the northern edge of the OS 1:25000 map 412-metre spot height. There are a series of prehistoric cairns on its slopes; Haytor Down is situated at the southeastern end of its greater summit area; Trendlebere Down and the East Dartmoor Woods and Heaths National Nature Reserve are located on its east-northeastern flank; there is a fair amount of woodland on its lower northern slopes, at the foot of which are Becky Falls, and Smallacombe Rocks and one or two Bronze Age stone hut circles are sited on its west-southwestern flank.

123=. Heatree Down, 1,355.97 feet/413.30 metres, SX 724802, a small, somewhat elongated, roundish-topped peak, which rises fairly steadily from the head of the Hayne Valley to the east and the East Webburn Valley to the west and variably from one of the tributary dales of the latter to the south and another of the River Bovey (via Vogwell Down) to the north. OS Maps shows its summit to be approximately 35 yards south of the OS 1:25000 map 414-metre spot height, which is situated on the top of a prehistoric cairn. There is a fair amount of woodland on one or two of its lower slopes.

123=. Stalldown Barrow, 1,355.97 feet/413.30 metres, SX 636622, a wedge-shaped, pretty rocky, flattish-topped peak, which rises steeply and then very gently from the Erme Valley to the east, variably from the same dale to the north and a col to the south and fairly steadily from the Yealm Valley to the west. OS Maps surprisingly computes its highest point to be approximately 35 yards west-southwest of the OS 1:25000 map 415-metre spot height, which is situated on the more easterly of two Bronze Age summit cairns, on the top of which are the remains of Hillson's House (perhaps a former peat cutter's shelter). There are numerous prehistoric remains on its slopes, including the impressive Stalldown (store) Row, enclosed settlements, cairns, cists and a cairn circle. A water treatment works is situated on its south-southwestern flank and there is a fair amount of woodland on its lower western slopes.

125. [Longstone Hill], 1,352.69 feet/412.30 metres, SX 567910, a small peak on a narrow, north-projecting spur of High Willhays, just over 1¼ miles to the south-southeast. It rises steeply from Meldon Reservoir to the west, steeply and then fairly steadily from a tributary dale of the Red-a-ven Brook to the east and variably from the Red-a-ven Valley to the north. OS Maps shows its summit to be at the OS 1:25000 map 412-metre spot height, but the land drops by only 5.91 feet to a very high col linking it to its parent hill.

126. [Hedge Down], 1,352.03 feet/412.10 metres, SX 735783, a somewhat elongated, partly rock, twin-summited peak on an east to north-curving spur of Chinkwell Tor, to the west-southwest. It rises variably from a tributary dale of the East Webburn River to the north and the Becka Valley to the east and quite gently from a tributary dale of the latter to the south. OS Maps surprisingly shows its summit to be at the OS 1:25000 map 413-metre spot height on its south-southeastern peak. Its north-northwestern top is located approximately 275 yards away, on a rocky outcrop at the OS 1:25000 map 414-metre spot height, and is computed by OS Maps to reach 1,345.47 feet. However, the hill doesn't have a relative height of fifty feet and is linked to its parent peak by a high col.

127=. [Bagtor Down], 1,349.74 feet/411.40 metres, SX 754751, a small, slightly elongated peak on a southeast-projecting spur of Rippon Tor. It rises fairly steeply from the Sig Valley to the east and variably from the same dale to the north and the Langworthy Valley (via Mountsland Common) to the south. OS Maps shows its summit to be approximately on top of the letter "n", in the word "Common", as printed on the OS 1:25000 map within the 410-metre small contour ring. However, the land drops by only 8.86 feet to a very high col linking it to its parent peak. A Bronze Age settlement is situated on its southern flank and several prehistoric cairns are located on its slopes. Horridge Common covers most of its upper slopes and there is a fair amount of woodland on its lower ones.

127=. [Middle Tor], 1,349.74 feet/411.40 metres, SX 669858, a somewhat elongated, partly rocky, flattish-topped peak on the upper eastern flank of Shovel Down, to the west. It is more immediately an outlier of nearby Kestor Rock, to the north-northwest. It rises variably from the South Teign Valley to the east and fairly steadily from one of its tributary dales to the south. OS Maps shows its rocky summit to be near the southeastern end of the OS 1:25000 map 410-metre contour ring, but the land drops by only 8.53 feet to a very high col linking it to its parent peak. Frenchbeer Rock is situated on its south-southeastern flank and there is a fair amount of woodland on its eastern and southeastern ones.

129. Sharpitor, 1,347.11 feet/410.60 metres, SX 559703, an impressive, wedge-shaped, pretty rocky peak, which rises increasingly steeply from a col to the north, fairly steadily from the Meavy Valley to the east and variably from Burrator Reservoir to the south and the Walkham Valley to the west. OS Maps shows its highest point to be on the top of its summit tor, within the OS 1:25000 map 410-metre very small contour ring. There are a series of Bronze Age remains on its slopes, including settlements, stone rows, stone hut circles and cairns, and there is a considerable amount of woodland on its lower ones. Leather Tor is situated on its south-southeastern flank; Peak Hill is located on its southwestern one and the villages of Walkhampton and Horrabridge are sited on its west-southwestern one. The B 3212 runs across its upper western and northern slopes, broadly from southwest to northeast.

130. Yar Tor, 1,345.47 feet/410.10 metres, SX 678739, an elongated, partly rocky peak, which rises steeply from the East Dart Valley to the west, fairly gently from one of its tributary dales to the north and gently from cols to the south-southeast and east. OS Maps shows its highest point to be on the top of its summit cairn, at the OS 1:25000 map 416-metre spot height. The fair discrepancy in the peak's altitude given by the online and paper versions of the OS's data might be explained by a limitation of the website's programme to calculate accurately the heights of rapidly-rising features. There are a series of Bronze Age remains on its flanks, including a triple stone row to the east and several cairns and stone hut circles. Yartor Down is situated on its southern flank and the remains of Dartmeet (clapper) Bridge are located at the foot of is south-southwestern one.

131. [Hartland Tor], 1,342.85 feet/409.30 metres, SX 641799, a small, pretty rocky peak on a south-projecting spur of White Ridge (and, more immediately, Stannon Tor), approaching 1½ miles to the north-northeast. It rises fairly steeply from the East Dart Valley to the west and south and very gently and then quite gently from the Stannon Valley to the east. OS Maps shows its summit to be approximately in the middle of the OS 1:25000 map 410-metre very small contour ring. However, it doesn't have a relative height of fifty feet and is linked to its parent peak by a high col. Several Bronze Age stone hut circles are found on its flanks and the hamlet of Postbridge is sited at the foot of its south-southeastern one.

132. [Haytor Down], 1,342.19 feet/409.10 metres, SX 765781, a small, flattish-topped peak towards the southeastern end of the curving plateau of Black

Hill. It rises steeply and then very gently from the Becka Valley to the west, pretty steeply from the head of a tributary dale of Reddaford Water to the east, fairly steadily from a slight depression to the north and gently from the Lemon Valley to the south. OS Maps shows its summit to be approximately 35 yards east of the OS 1:25000 map 407-metre spot height. However, the land drops by only 6.56 feet to a very high col linking it to its parent peak, to the northwest. The hamlet of Haytor Vale is situated on its south-southeastern flank; Smallacombe Rocks and one or two Bronze Age stone hut circles are located on its western one; the source of the River Lemon is sited at the foot of its south-southwestern one and its lower eastern slopes are covered by Yarner Wood. The B 3387 runs across its southeastern slopes, broadly from west-southwest to east-northeast.

133. [Brent Fore Hill], 1,332.02 feet/406.00 metres, SX 663620, an elongated, pretty rocky, flattish-topped peak on an east-southeast-projecting spur of Ugborough Hill. It rises fairly steadily from the East Glaze Valley to the south, fairly gently from a dry dale to the west, gently and then fairly gently from the Red Valley to the north and variably from a col to the south-southeast and the Avon Valley to the east. OS Maps shows its summit to be approximately 35 yards south of the OS 1:25000 map 408-metre spot height. However, it doesn't have a relative height of fifty feet and is linked to its parent peak by a high col. There are a number of prehistoric remains on its slopes and several small areas of woodland on its lower east-southeastern ones. Hickley Ridge is situated on its east-northeastern flank and Hickley Plain on its north-northeastern one.

134. [Shilstone Hill], 1,330.38 feet/ 405.50 metres, SS 761458, a flat-topped peak on a west-northwest-projecting spur of Exe Plain (and, more immediately, Hoar Tor and Hoccombe Hill), 2¼ miles to the south-southwest. It rises steeply and then gently from the Farley Valley (via Farley Hill) to the west, fairly steadily and then very gently from a tributary dale of Badgworthy Water (via Little Black Hill) to the east and variably from the East Lyn Valley to the north. OS Maps shows its summit to be approximately 150 yards east-southeast of the OS 1:25000 map trig point, which is marked as standing at 405 metres. However, the land drops by only 7.55 feet to a very high col linking it to its parent peak. Little Hill is situated on its southwestern flank; Scob Hill is located on its northwestern one and Tippacott Ridge, Malmsmead Hill and Cloud Hill are sited on its east-northeastern one. There are a number of areas of woodland on its lower slopes and the B 3223 runs across its western ones, broadly from south-southeast to north-northwest.

135. [Scorhill Down], 1,325.13 feet/403.90 metres, SX 657874, a somewhat elongated, roundish-topped peak, which rises quite steeply and then fairly gently from the North Teign Valley to the east, quite steeply and then gently from the same dale to the south and very gently and then quite gently from the Gallaven Valley to the west. OS Maps shows its summit to be at the OS 1:25000 map 405-metre spot height. However, it fails to be an independent hill by just 2.76 feet and instead is an outlier of Buttern Hill, to the north-northwest, to which it's linked by a high col. Scorhill Tor is situated on its southern flank; Teign-e-ver

Clapper Bridge is located at the foot of its south-southwestern one; the remains of a hut circle and the impressive Bronze Age Scorhill (stone) Circle) are sited on its west-southwestern one; Gidleigh Common covers most of its west-northwestern slopes and Gidleigh Tor is found on its east-northeastern flank. There is a fair amount of woodland on the lower slopes of its eastern hemisphere.

136. Wittaburrow, 1,315.94 feet/401.10 metres, SX 733752, an elongated, flat-topped peak, which rises variably from the Ruddycleave Valley to the south (via Pudsham Down), fairly steadily from the same dale to the east and the East Webburn Valley to the west and gently from a col to the north. OS Maps shows its highest point to be on the top of its summit rocks at the OS 1:25000 map 403-metre spot height. A sizable prehistoric cairn is situated at the southern end of its summit area and there is a considerable amount of woodland on its lower south-southwestern, southwestern and western slopes.

137. Swell Tor, 1,314.96 feet/400.80 metres, SX 560733, an ovalish-shaped, pretty rocky and heavily-quarried peak, which rises quite steeply from the Walkham Valley to the west, variably from one of its tributary dales to the south and gently from a dry valley to the east. OS Maps shows its summit to be approximately in the middle of the OS 1:25000 400-metre very small contour ring. Granite from (the now disused) Swelltor Quarries on its upper southern flank was used to build the Thames Embankment and Vauxhall Bridge, in London, and to widen the old London Bridge. There is a considerable amount of woodland on its lower western slopes.

138=. [Holwell Tor], 1,310.04 feet/399.30 metres, SX 751775, an elongated, pretty rocky peak on a north-northwest-projecting spur of Haytor Rocks, to the southeast. It rises increasingly steeply from the Becka Valley to the west and north. OS Maps shows its summit to be at the OS 1:25000 map 402-metre spot height. However, it doesn't have a relative height of fifty feet and is linked to its parent peak by a high col. A Bronze Age stone hut circle settlement is situated on its western flank and there are one or two disused quarries on its slopes.

138=. [Hound Tor (Houndtor Down)], 1,310.04 feet/399.30 metres, SX 742789, a very rocky peak, which rises gently from a tributary dale of the East Webburn River to the west, variably from the Becka Valley to the east and fairly steadily from a tributary dale of the latter to the north. OS Maps shows its summit to be on the highest point of its spectacular tors. The very large discrepancy in the peak's altitude given by the online and paper versions of the OS's data might be explained by a limitation of the website's programme to calculate accurately the heights of rapidly-rising features. It fails to be an independent hill by only 7.68 feet and instead is an outlier of Chinkwell Tor (and, more immediately, Hedge Down), approaching a mile to the west-southwest, to which it's linked by a high col. A Bronze Age cairn and cist are situated on its south-southwestern flank; the remains of the medieval village of Hundatora are located on its east-southeastern one and there are one or two small areas of woodland on its lower slopes.

140. Gripper's Hill, 1,308.40 feet/398.80 metres, SX 685655, a wedge-shaped, slightly rocky, flattish-topped peak, which rises steeply and then very gently from the Dean Valley to the east, increasingly gently from the Avon Dam Reservoir to the west, quite gently and then gently from the head of the Small Valley to the south and very gently from a col to the north-northwest. OS Maps shows its summit to be approximately 100 yards west-northwest of the OS 1:25000 map 398-metre spot height. There are quite a few prehistoric remains on its slopes; Smallbrook Plains are situated on its southern flank; Harbourne Head and Skerraton Down are located on its east-southeastern one and Lambs Down is sited on its east-northeastern one.

141. [Peek Hill], 1,299.21 feet/396.00 metres, SX 556699, a small, partly rocky peak, with a small tor, on a southwest-projecting spur of Sharpitor. It rises fairly steadily from a depression to the east, quite gently and then steeply from Burrator Reservoir to the south and quite gently and then quite steeply from the Black Valley to the west. OS Maps shows its summit to be just west of the OS 1:25000 map 400-metre spot height, by a Bronze Age cairn. The fair discrepancy in the peak's altitude given by the online and paper versions of the OS's data might be explained by a limitation of the website's programme to calculate accurately the heights of rapidly-rising features. It doesn't have a relative height of fifty feet and is linked to its parent peak by a very high col. There is a considerable amount of woodland on its southern slopes and the B 3212 runs across its western flank, broadly from south-southwest to north-northeast.

142. [Holwell Down], 1,295.28 feet/394.80 metres, SX 739776, a somewhat elongated peak on a very broad, east-southeast-projecting spur of Chinkwell Tor (and, more immediately, Bell Tor and Bonehill Down). It rises variably from the Becka Valley to the east, gently and then fairly steadily from one of its headstreams to the north and quite gently and then very gently from another to the south. OS Maps shows its summit to be at the OS 1:25000 map 399-metre spot height. The fair discrepancy in the peak's altitude given by the online and paper versions of the OS's data might be explained by a limitation of the website's programme to calculate accurately the heights of rapidly-rising features. It doesn't have a relative height of fifty feet and is linked to its parent peak by a high col. There are several small areas of woodland on its lower slopes.

143. Hayne Down, 1,291.34 feet/393.60 metres, SX 741803, an elongated, rocky, twin-summited peak, which rises steadily and then quite gently from a tributary dale of the Becka Brook to the south, increasingly steeply from the Hayne Valley to the east, quite gently and then fairly steeply from the same dale to the north, fairly gently from one of its tributary valleys to the west and quite gently from a col to the southwest. OS Maps shows its highest point to be on its northwestern summit, at the southwestern end of the OS 1:25000 map 390-metre very small contour ring, approximately 380 yards northwest of the 397-metre spot height. Its subsidiary (southeastern) top is about 25 yards south of the same spot height and computed by OS Maps to be just 2.63 feet lower. The fair discrepancy in the latter peak's altitude given by the online and paper versions of the OS's data might be explained by a limitation of the website's programme to calculate accurately the heights of rapidly-rising features. Bowerman's Nose, a tall granite stack, is located on the hill's northern flank and there is a fair amount of woodland on its lower eastern slopes.

144. [King's Tor], 1,287.40 feet/392.40 metres, SX 556738, a small, roundish, very rocky peak, which rises steeply and then steadily from the Walkham Valley (via Hucken Tor) to the west, quite gently and then fairly steadily from one of its tributary dales to the north and very gently and then gently from the head of another to the east. OS Maps shows its highest point to be on the top of its summit tor, within the OS 1:25000 map 400-metre tiny contour ring. The considerable discrepancy in the peak's altitude given by the online and paper versions of the OS's data might be explained by a limitation of the website's programme to calculate accurately the heights of rapidly-rising features. It fails by 9.98 feet to be an independent hill and instead is an outlier of Swell Tor to the south-southeast, to which it's linked by a high col. There is a fair amount of woodland on its lower western slopes.

145. White Hill, 1,280.18 feet/390.20 metres, SX 533838, a somewhat elongated, roundish-topped peak, which rises increasingly gently from the Walla Valley to the north, pretty gently from a tributary dale of the River Lyd (via Willsworthy Camp) to the west, gently from the Willsworthy Valley to the east, fairly gently from the last named to the south and very gently from a col to the east-northeast. OS Maps shows its summit to be at the OS 1:25000 map 390-metre spot height. A series of rifle ranges are situated on its southwestern flank and several prehistoric cairns on its west-southwestern one.

146. [Bonehill Down], 1,279.53 feet/390.00 metres, SX 735777, a small, flattish-topped peak protruding from the saddle between Chinkwell Tor and Holwell Down. It rises very gently from a tributary valley of the Becka Brook to the north and the heads of others of the East Webburn River to the west-southwest. OS Maps shows its summit to be approximately in the middle of the very small OS 1:25000 map 390-metre contour ring, but the land drops by only 5.25 feet to a very high col linking it to Holwell Down just to the east-southeast. As the latter is in itself an outlier of Chinkwell Tor (and, more immediately, Bell Tor), so is Bonehill Down. Bonehill Rocks are situated on its west-southwestern flank.

147. [Malmsmead Hill], 1,278.87 feet/389.80 metres, SS 786464, a small, flat-topped peak on a north-northwest-projecting spur of Exe Plain (and, more immediately, Hoar Tor, Hoccombe Hill and Shilstone Hill), well over 3¼ miles to the south-southwest. It rises steeply and then very gently from Lank Combe (via Great Black Hill) to the south and the Badgworthy Valley (via Cloud Hill) to the east, increasingly gently from Wat Combe to the north and quite gently and then very gently from the head of a tributary dale of the East Lyn River to the west. OS Maps shows its summit to be just to the east-southeast of the OS 1:25000 map 389-metre spot height. However, it doesn't have a relative height of fifty feet and is linked to its parent peak by a very high col. Little Black Hill is situated on its south-

southwestern flank; Southern Ball is located on its northern one and there is a fair amount of woodland on some of its lower slopes.

148. Leeden Tor, 1,276.90 feet/389.20 metres, SX 562716, an elongated, partly rocky, twin-summited peak, which rises fairly steadily from the Walkham Valley to the west, very gently and then fairly steadily from one of its tributary dales to the north and quite gently from a col to the south-southwest and a tributary valley of the River Meavy to the east. OS Maps shows its highest point to be at the OS 1:25000 map 389-metre spot height on its northern summit, while the main tor is situated on the northeastern edge of its summit area. Its subsidiary (southern top), approximately 475 yards away, reaches a height of 1,257.55 feet just south of the 385-metre spot height, according to OS Maps. A series of Bronze Age settlements, cairns and cists are situated on the hill's flanks; Ingra Tor is located on its west-northwestern one; there is a fair amount of woodland on its lower western slopes and the B 3212 runs across its lower southern and southeastern ones, broadly from south-southwest to north-northeast.

149. Meldon Hill, 1,276.25 feet/389.00 metres, SX 695861, an ovalish-shaped, partly rocky peak, which rises fairly steadily from a tributary dale of the South Teign River to the west, very gently and then quite steeply from the Teign Valley (via Meldon Common) to the north and variably from the West Bovey Valley to the south and the head of a tributary dale of the River Bovey to the east. OS Maps shows its summit to be just east-northeast of the OS 1:25000 map trig point, which is marked as standing at 390 metres. Padley Common and Chagford are situated on its north-northeastern flank and there are a number of small areas of woodland on its slopes.

150. Cripdon Down, 1,275.26 feet/388.70 metres, SX 734799, a wedge-shaped peak, which rises gently from a col to the west, fairly gently from another to the east, very gently from the East Webburn Valley (via Swine Down) to the south-southeast and variably from the Hayne Valley to the north. OS Maps shows its summit to be approximately 100 yards south of the OS 1:25000 map 398-metre spot height. There are a number of small areas of woodland on its slopes.

151. [Pudsham Down], 1,271.33 feet/387.50 metres, SX 732748, a small peak on a south-southwest-projecting spur of Wittaburrow, to the north. It rises steeply and then gently from the Ruddycleave Valley to the east, quite steeply and then gently from the same dale to the south and fairly steadily from the East Webburn Valley to the west. OS Maps shows its summit to be just north of the numbers "3" and "8" in the 387-metre spot height, as printed on the OS 1:25000 map. However, the land drops by only 5.91 feet to a very high col linking it to its parent peak. There is a considerable amount of woodland on its lower south-southwestern, southwestern and western slopes.

152. [Watchet Hill], 1,269.69 feet/387.00 metres, SX 614930, a small, flattish-topped peak on a north-projecting spur of Belstone Tor. It rises pretty steeply and then gently from the East Okement Valley to the west and variably from one of its tributary dales to the north and the Taw Valley to the east. OS Maps shows its summit to be situated approximately in the middle of the widest part of the spur, but the land drops by only 6.57 feet to a very high col linking it to its parent peak. A Bronze Age round cairn is situated at the north-northeastern end of its summit area and the village of Belstone and Brenamoor Common are located on its northeastern flank.

153. [Hart Tor], 1,268.70 feet/386.70 metres, SX 581719, a fairly rocky, roundish-topped peak on a south-southwest-projecting spur of South Hessary Tor, well over a mile to the east-northeast. It rises fairly steadily from the Meavy Valley to the west and very gently from the Hart Tor Valley to the south and east. OS Maps surprisingly shows its highest point to be just southeast of the OS 1:25000 map 390-metre tiny contour ring, within which is its summit tor. However, it doesn't have a relative height of fifty feet and is linked to its parent peak by a high col. A number of Bronze Age remains (a double and a single stone row and several cairns) are situated on its west-southwestern flank.

154. Buckland Beacon, 1,267.06 feet/386.20 metres, SX 735731, an elongated peak, which rises quite steeply from the Ruddycleave Valley to the west, very gently and then quite steeply from a col to the south-southeast, very gently from another to the north-northeast and variably from the Ashburn Valley to the east and the Dart Valley to the south-southwest. OS Maps shows its summit to be just to the east of where the name "Buckland Beacon" is printed on the OS 1:25000 map, approximately halfway between the two words and towards the southwestern end of the 380-metre contour ring. The two Ten Commandments Stones are situated just below Beacon Rock, near the top of the southwestern flank of the hill, and Welstor Rock is located on its upper east-southeastern one. There is a fair amount of woodland on its slopes and Buckland Common is sited on its northern flank.

155. [Leather Tor], 1,255.91 feet/382.80 metres, SX 562700, an impressive, very rocky peak on a short, south-projecting spur of Sharpitor, to the northwest. It rises fairly steadily and then quite steeply from the Meavy Valley to the east and gently and then quite steeply from Burrator Reservoir to the south and a depression to the west. OS Maps shows its highest point to be on the top of its dramatic, narrow summit tor, approximately in the middle of the OS 1:25000 map 380-metre small contour ring. However, the hill doesn't have a relative height of fifty feet and is linked to its parent peak by a high col. A Bronze Age cairn and cist are situated on its southern flank; the remains of a blowing house are located on its south-southeastern one and most of its lower slopes are covered by woodland.

156. [Zeal Hill], 1,252.62 feet/381.80 metres, SX 673639, a partly rocky, flattish-topped peak on a broad, south-projecting spur of Quickbeam Hill, approaching 1¾ miles to the west-northwest. It rises steeply and then gently from the Avon Valley to the east, quite steeply and then gently from the Bala Valley to the south and gently from the latter to the west. OS Maps shows its summit to be on the northern edge of the OS 1:25000 map 382-metre spot height, but the land drops by only 6.89 feet to a very high col linking it to its parent peak. Brent Moor covers most of its slopes, on which there are a number of prehistoric remains; the Hunters' Stone is situated on its southeastern flank and Black Tor is located on its east-southeastern one.

157. (Blackaton Hill), 1,249.67 feet/380.90 metres, SX 694785, an elongated and slightly rocky peak, which rises pretty steeply from the Broadaford Valley to the east, increasingly steeply from the West Webburn Valley to the west, variably from one of its tributary dales to the north and gently from a col to the south. OS Maps shows its summit to be at the OS 1:25000 map 381-metre spot height. The hill is unnamed on the OS 1:25000 and 1:50000 OS maps, but it seems appropriate to name it after Lower Blackaton and Blackaton Manor, which are situated on its southern slopes. One or two Bronze Age stone hut circles are located on its northern flank.

158. Riddon Ridge, 1,248.36 feet/380.50 metres, SX 666763, a small, elongated, slightly rocky plateau, which rises quite steeply and then very gently from the East Dart Valley to the west, fairly steadily and then gently from the same dale to the south, very gently from a col to the north and variably from the Walla Valley to the east. OS Maps shows its summit to be approximately in the middle of the OS 1:25000 map 380-metre very small contour ring. A series of Bronze Age stone hut circles and a cairn are situated on its slopes; the ruins of Whiteslade Farm (Snaily House) and a fair-sized wood are located on its west-southwestern flank and Bellever Clapper Bridge is sited at the foot of its northwestern one.

159. [Bonehill Rocks], 1,247.05 feet/380.10 metres, SX 731774, a small peak, with a mass of rock outcrops, on a short, west-projecting spur of Chinkwell Tor (and, more immediately, Bonehill Down and Holwell Down), to the north-northwest. It rises gently and then quite steeply from the East Webburn Valley to the west and its immediate tributaries to the south and north. OS Maps shows its summit to be at the OS 1:25000 map 393-metre spot height, but the land drops by only 5.91 feet to a very high col linking it to its parent peak. The very large discrepancy in the peak's altitude given by the online and paper versions of the OS's data might be explained by a limitation of the website's programme to calculate accurately the heights of rapidly-rising features.

160. Round Hill, 1,238.85 feet/377.60 metres, SS 806307, an elongated, roundish-topped peak, which rises gently and then very gently from Dane's Valley (via Black Ball) to the east, variably from tributary dales of the River Yeo to the south and west (the latter via Cussacombe Common) and increasingly gently from one of the latter valleys to the north. OS Maps shows its summit to be approximately on the north-northwest/south-southeast field boundary, just southeast of the OS 1:25000 map 378-metre spot height number. Molland Common and White Moor are situated on its east-southeastern flank; Long Breach is located on its east-northeastern one and there are narrow lines of woodland on some of the lowest slopes of its western hemisphere.

161. (Grendon Hill), 1,231.63 feet/375.40 metres, SX 681783, an oval-shaped peak, which rises variably from the West Webburn Valley to the east and one of its tributary dales to the north, gently from another to the south and very gently from the Walla Valley to the west. OS Maps shows its summit to be at the OS 1:25000 map 376-metre spot height. The hill is unnamed on the OS 1:25000 and 1:50000 maps, but it seems appropriate to name it after Grendon Cottage and Grendon Farm, which are situated on its slopes nearby. There are several strips of woodland on its flanks and Cator Common and the prehistoric Cator Pound are located on its west-southwestern one.

162. Ugborough Beacon, 1,228.67 feet/374.50 metres, SX 667591, a wedge-shaped, partly rocky, roundish-topped peak, which rises fairly steadily and then quite steeply from the Glaze Valley to the east, gently from the Lud Valley to the west and variably from the Scad Valley to the north and the Forder Valley and the village of Bittaford (via Moorhaven Village and Wrangaton Golf Course) to the south. OS Maps shows its highest point to be just northeast of its summit cairn, on which the OS 1:25000 map 378-metre spot height is marked. Beacon Rocks are situated at the east-northeastern end of its summit area; Eastern Beacon is one of three prehistoric cairns located on its greater summit area; Creber's Rock is sited on its south-southwestern flank and Beacon Plain is found on its west-northwestern one. There are quite a number of patches of woodland on its lower slopes. A railway line and the A 38 run across its lower southern and eastern slopes, broadly from west-southwest to east-northeast.

163. [Wind Tor (Dunstone Down)], 1,228.35 feet/374.40 metres, SX 707758, an elongated, partly rocky peak on a south-projecting spur of Hamel Down (and, more immediately, Hameldown Beacon), well over 2½ miles to the north. It rises gently and then quite gently from a tributary valley of the West Webburn River to the west and variably from the West Webburn Valley (via Bittleford Down) to the south and the East Webburn Valley to the east. OS Maps shows its summit to be approximately 35 yards northeast of the last digit in the word "Tor", as marked on the OS 1:25000 map. However, the hill doesn't have a relative height of fifty feet and is linked to its parent peak by a very high col. Two Bronze Age stone hut circles are situated on its upper slopes and most of the lowest ones of its southern hemisphere are wooded.

164. South Down, 1,227.69 feet/374.20 metres, SX 555912, an oval-shaped, slightly rocky, flattish-topped peak, which rises steeply and then fairly steadily from Meldon Reservoir to the east and south, quite gently and then fairly steadily from the Lew Valley to the west and variably from a tributary dale of the West Okement River to the north. OS Maps shows its summit to be on the southern edge of the OS 1:25000 map 375-metre spot height. Meldon Viaduct and a line of woodland are situated at the foot of its north-northeastern flank and the A 30 cuts across its east-northeastern slopes, broadly from west-southwest to east-northeast.

165. [Black Tor], 1,220.47 feet/372.00 metres, SX 573718, a partly rocky, flattish-topped peak on a south-projecting spur of North Hessary Tor, approaching

1½ miles to the north-northeast. It rises fairly steadily from the Meavy Valley to the east, fairly steadily and then quite gently from the same dale to the south and very gently from a col to the west. OS Maps shows its highest point to be on its summit tor, towards the south of the OS 1:25000 map 370-metre small contour ring. However, it doesn't have a relative height of fifty feet and is linked to its parent peak by a very high col. A logan stone is situated at its summit; its main tor is located on its southeastern flank, at the foot of which are the remains of a blowing house, and Bronze Age settlements and cairns and a double stone row are sited on its south-southwestern slopes. The B 3212 runs across its western flank, broadly from south-southwest to north-northeast.

166. [Bagga Tor], 1,214.57 feet/370.20 metres, SX 548805, a small, very rocky peak on a west-northwest-projecting spur of Cut Hill (and, more immediately, Walkham Head and Lynch Tor), approaching 3½ miles to the east-northeast. It rises quite steeply and then fairly steadily from the Tavy Valley to the west, increasingly steeply from the Baggator Valley to the north and fairly steadily and then steadily from the Youldon Valley to the south. OS Maps shows its highest point to be on its summit tor, at the OS 1:25000 map 372-metre spot height. However, it doesn't have a relative height of fifty feet and is linked to its parent peak by a very high col. There are one or two small areas of woodland on its slopes.

167. Sharp Tor, 1,214.24 feet/370.10 metres, SX 686729, a wedge-shaped, very rocky, pointed peak, which rises increasingly steeply from a col to the north-northwest and the head of a tributary dale of the River Dart to the west, fairly steadily from Simon's Lake to the east and variably from the Dart Valley (via a small wood) to the south. OS Maps surprisingly shows its summit to be just southeast of the OS 1:25000 map 380-metre tiny contour ring and summit rocks. The very large discrepancy in the peak's altitude given by the online and paper versions of the OS's data might be explained by a limitation of the website's programme to calculate accurately the heights of rapidly-rising features. Several Bronze Age stone hut circles are situated on its flanks and Luckey Tor is located near the foot of its south-southwestern one.

168. [Yartor Down], 1,206.36 feet/367.70 metres, SX 679730, a roundish, slightly rocky peak, which rises steeply and then gently from the Dart Valley to the south, fairly steeply and then quite steadily from the same dale to the west, quite steeply and then gently from a dry valley to the east and very gently from a col to the north-northeast. OS Maps shows its summit to be at the OS 1:25000 map 368-metre spot height, but it fails to be an independent hill by just 4.40 feet and instead is an outlier of Yar Tor, to the north, to which it's linked by a high col. There are quite a number of Bronze Age remains on its slopes and the Coffin Stone (a former resting site of bodies in transit) is situated at the foot of its northwestern flank.

169. Sheeps Tor (Yellowmead Down), 1,204.07 feet/367.00 metres, SX 565682, an impressive, elliptical-shaped, very rocky peak, which rises fairly steadily and then steeply from Burrator Reservoir to the north and west and the Sheepstor Valley to the south and very gently and then quite gently from a tributary dale of the Narrator Brook to the east. OS Maps shows its highest point to be on the top of its summit tor, at the OS 1:25000 map 369-metre spot height. The Piskies House cave is situated on its south-southeastern flank; a number of pillow mounds are located on its western one and its lower western, northern and east-northeastern ones are covered by woodland.

170. Rowley Down, 1,198.82 feet/365.40 metres, SS 670432, an elongated, flat-topped peak, which rises steeply and then very gently from a headwater valley of the River Heddon (via Holwell Rocks) to the east, quite gently and then very gently from one of the River Bray (via Higher Down) to the south and the head of a tributary dale of the River Heddon to the west and variably from the Heddon Valley to the north. OS Maps shows its highest point to be at the 365-metre OS 1:25000 map spot height. The Bronze Age Holwell Barrow is situated at the southeastern end of its summit area; other prehistoric remains are located on its slopes; Holwell Castle is sited on its lower northern flank; there are patches of woodland on some of its lower slopes and Wistlandpound Reservoir lies at the foot of its west-southwestern flank. Three important roads run across its lower slopes: the A 39 to the north and northwest, the A 399 to the west-southwest and the B 3358 to the south.

171. [Combshead Tor], 1,198.49 feet/365.30 metres, SX 587688, a small, pretty rocky peak on a south-projecting spur of Higher Hartor Tor, to the east-southeast. It rises fairly steadily from the Narrator Valley to the east, south and west. OS Maps shows its highest point to be on the top of its summit tor, at the OS 1:25000 map 371-metre spot height. The considerable discrepancy in the peak's altitude given by the online and paper versions of the OS's data might be explained by a limitation of the website's programme to calculate accurately the heights of rapidly-rising features. It doesn't have a relative height of fifty feet and is linked to its parent peak by a high col. The impressive Cuckoo Rock is situated on its west-southwestern flank and there are disused tin workings on its slopes.

172. [Butterdon Hill], 1,197.51 feet/365.00 metres, SX 655586, a somewhat elongated, partly rocky peak on a south-projecting spur of Ugborough Hill (and, more immediately, Sharp Tor and Piles Hill), approaching 2½ miles to the north. It rises fairly steadily and then quite gently from the Lud Valley to the east and variably from Black Pool at a col to the south and the Erme Valley to the west. OS Maps shows its summit to be just north of the OS 1:25000 map trig point, which is marked as standing at 364 metres. However, it fails to be an independent hill by just 5.38 feet and is linked to its parent peak by a high col. Numerous Bronze Age remains are situated on its slopes, including two stone rows, a long barrow, hut circles and several cairns. Weatherdon Hill, Butter Brook Reservoir, Tor Rocks and a disused quarry are located on its west-northwestern flank; Hangershell Rock is sited on its northern one and Main Head is lies on its northeastern one.

173. [Scob Hill], 1,197.18 feet/ 364.90 metres, SS 752466, a small, flattish-topped peak on a west-northwest-projecting spur of Exe Plain (and, more immediately, Hoar Tor, Hoccombe Hill and Shilstone Hill), approaching 2¾ miles to the south. It rises steeply and then gently from the East Lyn Valley to the north and one of its tributary dales to the east and quite steeply and then gently from a tributary valley of Farley Water to the west. OS Maps shows its summit to be approximately in the middle of the widest part of the OS 1:25000 map 360-metre contour bulge. However, the land drops by only 12.80 feet to a very high col linking it to its parent peak. There are one or two disused quarries on its slopes and almost all its lower ones are wooded. The B 3223 runs across its southern, western and west-northwestern flanks, broadly from south-southeast to north-northwest.

174. [Glasscombe Ball], 1,192.91 feet/363.60 metres, SX 658603, a small peak on a south-southeast-projecting spur of Ugborough Hill (and, more immediately, Sharp Tor and Piles Hill), approaching 1½ miles to the north-northwest. It rises fairly steadily from the Glaze Valley to the east, very gently from a col to the south and variably from the Erme Valley to the west. OS Maps shows its highest point to be approximately 80 yards west-northwest of the prehistoric cairn on its summit area, within the OS 1:25000 map 360-metre contour bulge. However, the land drops by only 9.17 feet to a very high col linking it to its parent peak. One or two prehistoric cairns are situated on its western flank and the remains of a blowing house on its lower eastern one.

175. [Badgworthy Hill], 1,187.99 feet/362.10 metres, SS 789437, a small, flattish-topped peak on an east to northeast-curving spur of Exe Plain (and, more immediately, Hoar Tor and Hoccombe Hill), well over 2¼ miles to the west-southwest. It rises quite steeply and then very gently from a tributary valley of Hoccombe Combe to the north and the Badgworthy Valley to the east and fairly steeply and then gently from Hoccombe Water to the south. OS Maps shows its summit to be at the OS 1:25000 map 364-metre spot height, but the land drops by only 5.58 feet to a very high col linking it to its parent peak. There are one or two prehistoric remains on its slopes.

176. [Scarey Tor], 1,187.34 feet/361.90 metres, SX 607924, a small, partly rocky peak on a west-northwest-projecting spur of Belstone Tor. It rises quite steeply and then fairly steadily from the East Okement Valley to the west and north and very gently from the same dale to the south. OS Maps shows its summit to be approximately 50 yards south of the OS 1:25000 map 365-metre spot height, but the land drops by only 8.53 feet to a very high col linking it to its parent peak. Cullever Steps river crossing and plunge pool are situated at the foot of its south-southwestern flank.

177. Down Tor, 1,183.73 feet/360.80 metres, SX 580693, a small, somewhat elongated, rocky peak, which rises fairly steadily from the Narrator Valley to the south and Newleycombe Lake to the north, fairly gently from a col to the east and variably from the Meavy Valley to the west. OS Maps shows its highest point to be on the top of its summit tor, at the OS 1:25000 map 366-metre spot height. The fair discrepancy in the peak's altitude given by the online and paper versions of the OS's data might be explained by a limitation of the website's programme to calculate accurately the heights of rapidly-rising features. A number of Bronze Age stone hut circles are situated on its slopes and there is a fair amount of woodland on its lower western and southwestern ones.

178. Brimblecombe Hill, 1,183.40 feet/360.70 metres, SS 833299, an elongated peak, which rises fairly steeply and then gently from a tributary valley of Dane's Brook to the north, very gently from a col to the west-northwest and variably from another to the east-southeast, Dane's Valley to the east and Ringcombe to the south. OS Maps shows its summit to be approximately in the middle of the OS 1:25000 map 360-metre contour ring. West Anstey Common, Guphill Common and Anstey Rhiney Moor are situated on its east-southeastern flank and there are one or two woods on its lower slopes.

179. [Greator Rocks], 1,177.82 feet/359.00 metres, SX 747786, a small, elongated, very rocky peak on an east-northeast-projecting spur of Hound Tor, to the northwest. It rises variably from the Becka Valley to the south, steeply from the same dale to the east and fairly steeply from one of its tributary dales to the north. OS Maps shows its highest point to be on the top of the more easterly of its two main tors. The very large discrepancy in the peak's altitude given by the online and paper versions of the OS's data might be explained by a limitation of the website's programme to calculate accurately the heights of rapidly-rising features. It doesn't have a relative height of fifty feet and is linked to its parent peak by a very high col. One or two Bronze Age stone hut circles are situated on its south-southeastern flank and its lower northern and northeastern slopes are heavily wooded.

180. [Cheriton Ridge], 1,174.54 feet/358.00 metres, SS 743451, a short, broad ridge on a north-northwest-projecting spur of Exe Plain, well over 1¾ miles to the south-southeast. It rises steeply and then gently from the Farley Valley to the east, steeply and then very gently from the Hoaroak Valley to the north and quite steeply and then gently from the latter to the west. OS Maps shows its summit to be at the southeastern edge of the OS 1:25000 map 359-metre spot height. However, it doesn't have a relative height of fifty feet and is linked to its parent peak by a high col. There is a fair amount of woodland on most of its lower slopes.

181. (Mardon Down West Top), 1,169.95 feet/356.60 metres, SX 767870, the western, and higher, summit of Mardon Down, a large semicircular area of upland, which rises variably from the Wray Valley to the west, one of its tributary dales to the south, the Teign Valley to the north and one of its tributaries to the east. OS Maps shows its highest point to be approximately 50 yards north of the OS 1:25000 map 356-metre spot height, towards the southern end of its very small, elongated, flat-topped summit area. The specific peak is unnamed on the OS 1:25000 and 1:50000 maps, except under the generic name of Mardon Down, but it seems appropriate to add West Top to its title because of its location in relation to the separate, eastern summit. A series of prehistoric remains are situated on its greater summit area, including the Giant's Grave, a stone circle and a cairn circle. There is a fair amount of woodland on its lower slopes and the B 3212 snakes across its lower southern flank, broadly from west-southwest to east-northeast.

182. (Venford Hill), 1,167.32 feet/355.80 metres, SS 856290, an elongated, flattish-topped peak, which is the most easterly 1,000-foot hill in Devon. It rises steeply and then gently from Dane's Valley to the north, fairly steadily and then quite gently from the head of a tributary valley of the River Yeo (via Woodland Common) to the south, gently from a col to the west-northwest and variably from the head of a tributary dale of the Brockey River to the east-southeast. OS Maps shows its summit to be at the 356-metre OS 1:25000 map trig point. The hill is unnamed on the OS 1:25000 and 1:50000 maps, but it seems appropriate to name it after Venford Moor, which covers most of its upper east-southeastern slopes. The Bronze Age West Anstey Barrows are situated towards the western end of its greater summit area; Twitchen Common is located on its southeastern flank; East Anstey Common, Liscombe Allotment, the Bronze Age Anstey Barrow and the source of the River Yeo are all sited on its east-southeastern one; Whiterocks Down lies on its east-northeastern one; Great Common is found on its north-northeastern one and Anstey Money Common is positioned on its north-northwestern one. There is a line of woodland along its lowest north-northeastern to eastern slopes.

183. [Combestone Tor], 1,165.03 feet/355.10 metres, SX 670718, a small, pretty rocky peak on a north-projecting spur of Ryder's Hill (and, more immediately, Holne Ridge), well over 1¾ miles to the south-southwest. It rises variably from the Dart Valley to the north and fairly steadily and then quite gently from one of its tributary dales to the east and the O Valley to the west. OS Maps shows its highest point to be on the top of its summit tor, just southeast of the OS 1:25000 map 356-metre spot height. However, it doesn't have a relative height of fifty feet and is linked to its parent peak by a very high col. A series of rock basins are situated on its summit area; several Bronze Age stone hut circles are located on its flanks and a line of woodland runs along its lowest northeastern slopes.

184. Black Down, 1,159.45 feet/353.40 metres, SX 522824, an elongated, flat-topped peak, which rises quite gently and then very gently from a tributary dale of the River Lyd to the north and variably from the Burn Valley to the west, the Tavy Valley (via Horndon Down and Wheal Jewell Reservoir) to the south and the Willsworthy Valley (via Wheal Jewell Leat) to the east. OS Maps shows its summit to be just west of the OS 1:25000 map 354-metre spot height. A pool is situated on its summit area, the village of Horndon on its southern flank and Kingsett Down on its south-southwestern one. There are some small areas of woodland on its lower slopes; the source of the River Burn is situated near the foot of its west-northwestern flank and the A 386 runs across its western one, broadly from south-southwest to north-northeast.

185. Rowden Down, 1,158.46 feet/353.10 metres, SX 698760, a somewhat elongated, flat-topped peak, which rises fairly steadily from the West Webburn Valley to the west, fairly steadily and then gently from the same dale to the south, variably from one of its tributary valleys to the north and gently from a col to the east. OS Maps shows its summit to be just north of the last two digits of the OS 1:25000 map 356-metre spot height number. A Bronze Age stone hut circle is situated on its south-southwestern flank.

186. [Southern Ball], 1,156.82 feet/352.60 metres, SS 784475, a small, flattish-topped peak on an east-northeast-projecting spur of Exe Plain (and, more immediately, Hoar Tor, Hoccombe Hill, Shilstone Hill and Malmsmead Hill), well over 3¾ miles to the south-southwest. It rises fairly steadily and then quite gently from the East Lyn Valley to the east and variably from the same dale to the north and the head of Wat Combe to the south. OS Maps shows its summit to be at the OS 1:25000 map 353-metre spot height. However, it doesn't have a relative height of fifty feet and is linked to its parent peak by a high col. Southern Wood covers most of its lower northern and northeastern slopes.

187. [Moorhouse Ridge], 1,154.53 feet/351.90 metres, SS 824311, a somewhat elongated, flat-topped peak on an east-southeast-projecting spur of North Molton Ridge (and, more immediately, Twitchen Ridge), approaching three miles to the west-northwest. It rises fairly steeply and then gently from Dane's Valley to the north, fairly steadily and then gently from the same dale to the east and gently from one of its tributary valleys to the south. OS Maps shows its summit to be on the southwestern end of the digit "3", in the 353-metre spot height number, as printed on the OS 1:25000 map. However, it doesn't have a relative height of fifty feet and is linked to its parent peak by a very high col. Soakey Moor is situated on its lower western flank and there is an area of woodland on its lowest east-southeastern one.

188. Gibbet Hill, 1,154.20 feet/351.80 metres, SX 503811, an elongated, roundish-topped peak, which rises steeply and then fairly gently from the Cholwell Valley to the east, fairly steadily from the village of Mary Tavy to the south and a tributary dale of the River Burn to the north, fairly steadily and then quite gently from the Burn Valley to the west and very gently and then gently from a col to the northeast. OS Maps shows its summit to be at the OS 1:25000 map trig point, which is marked as standing at 353 metres. The peak is named after a gibbet, which once stood on top of it. Blacknor Park is situated on its west-southwestern flank and the remains of the Wheal Betsy engine house, with its leaning chimney, are located on its east-northeastern one. The A 386 runs across its western and southern slopes, broadly from south-southwest to north-northeast.

189. Pepperdon Down, 1,153.22 feet/351.50 metres, SX 779852, a wedge-shaped, roundish-topped peak, which rises very gently from a col to the south, variably from the Wray Valley (via Steward Wood) to the west and the Trenchford Valley to the north and fairly steadily from the latter to the east. OS Maps shows its summit to be at the OS 1:25000 map 353-metre spot height. Hingston Down is situated on its west-northwestern flank and there is a fair amount of woodland on the lower slopes of its western hemisphere.

190=. [Bel Tor], 1,151.90 feet/351.10 metres, SX 694731, a small, partly rocky peak on an east-southeast-projecting spur of Corndon Tor, approaching a mile to the north-northwest. It rises steeply and then very gently from the Dart Valley to the south, fairly steadily and then gently from one of its tributary dales to the east and quite gently and then very gently from Simon's Lake to the west and a tributary valley of the West Webburn River to the north. OS Maps shows its summit to be approximately in the middle of the OS 1:25000 map 350-metre very small contour ring, around 160 yards west-northwest of its tor and trig point. However, the land drops by only 10.50 feet to a very high col linking it to its parent peak. A logan stone and a rock basin are situated on its tor; Mel Tor is located on its southern flank; there is a fair amount of woodland on its lower southern slopes and Aish Tor is sited on its south-southeastern flank.

190=. Butterdon Down (Butterdon Hill), 1,151.90 feet/351.10 metres, SX 750884, a round peak, which rises increasingly steeply from a col to the west, increasingly gently from another to the east-northeast, fairly gently from a tributary dale of the River Teign to the south and variably from the Teign Valley to the north. OS Maps shows its summit to be at the OS 1:25000 map 351-metre spot height, on the top of a prehistoric cairn. Willingstone Rock is situated on its north-northeastern flank; an impressive monolith is located on its western one; its lower northern slopes are heavily wooded and there are a number of smaller woods on its other ones.

192. Silkwood Top, 1,150.26 feet/350.60 metres, SS 700374, a wedge-shaped peak, which rises steeply and then very gently from the Bray Valley to the west, steeply and then gently from one of its tributary dales to the east, variably from the same valley to the south and increasingly gently from another to the north. OS Maps shows its summit to be approximately in the middle of the OS 1:25000 map 350-metre very small contour ring. There is a considerable amount of woodland on its lower eastern and western slopes.

193. East Hill, 1,147.31 feet/349.70 metres, SX 595938, an oval-shaped peak, which rises steeply and then quite steeply from the East Okement Valley to the north, variably from the same dale to the east-northeast, gently from a col to the west-southwest and very gently from the Moor Valley to the south. OS Maps shows its summit to be approximately in the middle of the OS 1:25000 map 350-metre very small contour ring, near to the trig point, which is marked as standing at 349 metres. An Iron Age hillfort is situated on its east-northeastern flank; there is a fair amount of woodland on its lower eastern slopes and the A 30 and the Dartmoor Railway run across its lower northern ones, broadly from west-southwest to east-northeast.

194=. Cosgate Hill, 1,146.33 feet/349.40 metres, SS 791488, an ovalish-shaped peak, which rises steeply from the East Lyn Valley to the south and The Combe to the north, steeply and then gently from Coscombe to the east and very gently from a col to the west-northwest. OS Maps shows its summit to be towards the eastern end of the OS 1:25000 map 340-metre contour ring. A series of Bronze Age barrows run across its greater summit area; most of its upper northern slopes and its lower southern ones are wooded and Ashton Cleave is situated at the foot of its southwestern flank. The A 39 curves round its eastern and northern slopes, broadly from east-southeast to west-northwest.

194=. [Dockwell Ridge], 1,146.33 feet/349.40 metres, SX 687637/SX 686640, an elongated, pretty rocky, flattish-topped, twin-summited peak on a south to south-southeast-curving spur of Gripper's Hill, to the north. It rises fairly steadily and then gently from the Avon Valley to the west and variably from one of its tributary dales to the south and the Harbourne Valley to the east. OS Maps computes its two summits to be of an identical altitude, the south-southeastern one being at the OS 1:25000 map 351-metre spot height and the north-northwestern one just over ¼ mile away, at the southeastern end of the map's 350-metre small contour ring. However, the hill doesn't have a relative height of fifty feet and is linked to its parent peak by a high col. A number of prehistoric remains are situated on its slopes; Shipley Tor is located on its south-southwestern flank; the Woolholes are sited on its west-northwestern one; disused tin workings are found on its northeastern one and there are one or two small areas of woodland on its lower slopes.

196. Ringmoor Down, 1,146.00 feet/349.30 metres, SX 575667, a wedge-shaped, slightly rocky peak, which rises fairly steadily from Gutter Mire to the east, fairly steadily and then very gently from Smallacombe to the west, quite gently and then gently from the Plym Valley to the south and variably from the Sheepstor Valley to the north. OS Maps shows its summit to be at the OS 1:25000 map trig point, which is marked as standing at 350 metres. Gutter Tor is situated at the east-northeastern end of its summit area; there are numerous prehistoric remains on its slopes, especially a Bronze Age stone row and cairn circles to the southwest; Legis Tor, Legistor Warren and disused tin workings are located on its south-southwestern flank; Legis Mire lies at the foot of its southwestern one; Lynch Common is sited on its west-southwestern one; Burrator Waterfall is found at the foot of its west-northwestern one and Meavy Pool can be seen at the bottom of its south-southeastern one. There are some small areas of woodland on its lower slopes.

197. [Great Trowlesworthy Tor], 1,143.37 feet/348.50 metres, SX 579643, an elongated, very rocky peak on a northwest-projecting spur of Shell Top, approaching 1½ miles to the east. It rises gently and then fairly steadily from the Blacka Valley to the west and gently from one of its headwater streams to the south and Spanish Lake to the north. OS Maps shows its highest point to be on the top of its summit tor, at the OS 1:25000 map 357-metre spot height. The sizable discrepancy in the peak's altitude given by the online and paper versions of the OS's data might be explained by a limitation of the website's programme to calculate accurately the heights of rapidly-rising features. It doesn't have a relative height of fifty feet and is linked to its parent peak by a high col. Numerous Bronze Age remains, including enclosed settlements, stone circles, stone rows and stone hut circles, and pillow mounds are situated on its slopes and Little Trowlesworthy Tor and Trowlesworthy Warren are located on its northwestern flank.

198. (Mardon Down East Top), 1,142.72 feet/348.30 metres, SX 776876, the eastern, and lower, summit of Mardon Down, a large semicircular area of upland. The East Top is somewhat elongated, slightly rocky and roundish-topped. It rises fairly steadily from a dry valley to the west, fairly steadily and then gently from a tributary dale of the River Teign to the south, increasingly gently from another of the latter (via Coleridge Wood) to the north and variably from yet another to the east. Because it has a relative height of 93.18 feet, this eastern top qualifies as a 1,000-foot peak in its own right. OS Maps shows its summit to be just northeast of the OS 1:25000 map 349-metre spot height. The specific peak is unnamed on the OS 1:25000 and 1:50000 maps, except under the generic name of Mardon Down, but it seems appropriate to add East Top to its title because of its location in relation to the separate, western summit. The Headless Cross monolith is situated on its northwestern flank and there is a considerable amount of woodland on its lower slopes.

199. Holdstone Hill (Holdstone Down), 1,139.44 feet/347.30 metres. SS 619477, an oval-shaped peak, which rises very steeply and then fairly steadily from the Bristol Channel (via sea cliffs) to the north, pretty steeply and then fairly steadily from Sherrycombe to the west, fairly gently from a col to the east and variably from a tributary valley of the River Heddon to the south. OS Maps shows its summit to be by a great modern rock cairn at the OS 1:25000 map trig point, which is marked as standing at 349 metres. Red Cleave is situated at the foot of its northern flank and there are a few small areas of woodland on its lower slopes.

200. [Gutter Tor], 1,137.14 feet/346.60 metres, SX 577668, a small, rocky peak at the east-northeastern end of the summit area of Ringmoor Down. It rises increasingly steeply from the Sheepstor Valley to the north, fairly steadily from Gutter Mire to the east and gently from the Plym Valley to the south. OS Maps shows its highest point to be on the top of its summit tor, at the east-northeastern end of the OS 1:25000 map 340-metre contour ring. However, the land drops by only 11.81 feet to a very high col linking it to its parent peak. There are several prehistoric remains on its slopes and Meavy Pool is situated at the foot of its southern flank.

201. Blackingstone Rock, 1,136.81 feet/346.50 metres, SX 786855, a very large and impressive tor at the western end of a short, south-southeast to west-running ridge. It rises increasingly steeply from the north and variably from the Trenchford Valley to the south, the Kennick Valley to the east and the head of the Beadon Valley to the west. OS Maps shows its summit to be at the OS 1:25000 map 355-metre spot height on the top of the rock, which is accessed on a stone staircase with handrails! The sizable discrepancy in the peak's altitude given by the online and paper versions of the OS's data might be explained by a limitation of the website's programme to calculate accurately the heights of rapidly-rising features. Most of its slopes are wooded.

202. [Great Combe Tor], 1,136.48 feet/346.40 metres, SX 525770, a small, partly rocky peak on a north-northwest-projecting spur of Cox Tor. It rises fairly steadily from the Colly Brook to the north, very gently from the head of one of its tributary dales to the east and gently and then fairly steadily from the Tavy Valley to the west. OS Maps shows its summit to be on the top of a rock outcrop approximately 60 yards south-southeast of the northern point of the OS 1:25000 map 340-metre contour bulge. However, the land drops by only 12.46 feet to a very high col linking it to its parent peak. Its main tor is situated on its north-northwestern flank and the village of Peter Tavy at the foot of its west-northwestern one.

203. [Old Burrow Hill], 1,132.87 feet/345.30 metres, SS 785490, an elongated peak, which rises very steeply and then very gently from the Bristol Channel (via Handball cliff) to the north, increasingly gently from Wingate Combe to the west and variably from the East Lyn Valley to the south. OS Maps shows its summit to be at the OS 1:25000 map 345-metre spot height. However, it fails to be an independent peak by only 6.37 feet and is instead is an outlier of nearby Cosgate Hill, to the east-southeast, to which it's linked by a high col. The remains of the Old Burrow Roman fortlet are situated on its northeastern flank; the Giant's Rib natural arch is located at its foot and there is a fair amount of woodland on its northern slopes. The A 39 runs across its southern flank, broadly from east to west.

204. Kipscombe Hill, 1,121.39 feet/341.80 metres, SS 765493, a roundish-shaped peak, which rises very steeply and then quite gently from the Bristol Channel and Countisbury Cove (via a sea cliff) to the north, quite gently from a col to the west-northwest, very gently from another to the east-southeast and variably from the East Lyn Valley to the south. OS Maps shows its summit to be just northeast of the OS 1:25000 map trig point, which is marked as standing at 343 metres. Countisbury Common covers most of its upper slopes and there is a fair amount of woodland on its northern ones, but rather less on its lower southern ones. The A 39 runs across its southern flank, broadly from east-southeast to west-northwest.

205. [Mel Tor], 1,115.16 feet/339.90 metres, SX 693725, a rocky peak on a south-projecting spur of Corndon Tor (and, more immediately, Bel Tor), well over a mile to the north-northwest. It rises steeply from Simon's Lake to the west and the Dart Valley (via Meltor Wood) to the south and fairly steadily from the head of one of the tributary dales of the latter to the east. OS Maps shows its highest point to be on the top of its summit tor, just east-southeast of the OS 1:25000 map 346-metre spot height. The considerable discrepancy in the peak's altitude given by the online and paper versions of the OS's data might be explained by a limitation of the website's programme to calculate accurately the heights of rapidly-rising features. It doesn't have a relative height of fifty feet and is linked to its parent peak by a high col. Two Bronze Age stone hut circles are situated on its east-southeastern flank; Hockinston Tor, Sharrah Pool Marsh and Aish Tor are located on its south-southeastern one and Mel Pool is sited at the foot of its southern one.

206. Kentisbury Down, 1,106.96 feet/337.40 metres, SS 639433, an isolated, somewhat elongated, roundish-topped peak, which rises fairly steeply and then quite gently from tributary valleys of the River Heddon to the east and the River Yeo to the south and variably from the former to the north and another tributary dale of the latter to the west. OS Maps shows its summit to be approximately 125 yards south of the OS 1:25000 map trig point, which is marked as standing at 337 metres. An Iron Age hillfort is situated on its east-southeastern flank; a dispersed group of standing stones are located on its north-northwestern one; the hamlet of Kentisbury is sited on its west-northwestern one and Cowley Wood is found on its lower north-northeastern slopes. The A 39 crosses its southern flank, broadly from west-southwest to east-northeast, and the A 399 curves across its eastern and northern ones, essentially from southeast to northwest.

207. Cranbrook Down (Uppacott Down), 1,104.99 feet/336.80 metres, SX 738889, a wedge-shaped, partly rocky, round-topped peak, which rises steeply and then quite gently from the Teign Valley to the north, quite gently from one of its tributary dales to the east and variably from another to the south and west. OS Maps shows its summit to be approximately in the middle of the OS 1:25000 map 330-metre contour ring, within the ramparts of Cranbrook Castle, an Iron Age hillfort. There is a considerable amount of woodland on its northern and eastern slopes.

208. Lyn Down, 1,104.66 feet/336.70 metres, SS 725465, an elongated, twin-summited peak, which rises steeply and then gently from the Hoaroak Valley to the east, increasingly gently from the West Lyn Valley to the west, very gently from a col to the south and variably from the East Lyn Valley (via Lyn Cleave and Summer House Hill) to the north. OS Maps shows its main (southern) summit to be at the OS 1:25000 337-metre spot height and its subsidiary (northern) top to be a little west of the 330-metre contour ring, approximately 1,100 yards to the north, and 13.45 feet lower. The prehistoric Lyn Long Stones are situated on its northern summit area; Stock Castle Iron Age enclosure is located on its western flank; the village of Barbrook is sited at the foot of its west-northwestern one; Glen Lyn Gorge can be seen at the bottom of its north-northwestern one and Shortacombe Common is found on its eastern one. There is a line of woodland along most of its lowest slopes.

209. [Barn Hill], 1,100.72 feet/335.50 metres, SX 533747, a partly rocky, flattish-topped peak on a south-southeast-projecting spur of Cox Tor, approaching a mile to the north-northwest. It rises fairly steadily from a tributary valley of the River Walkham to the east, fairly steadily and then gently from one of the River Tavy to the west and gently from a col to the south-southeast. OS Maps shows its summit to be at the OS 1:25000 map 336-metre spot height. However, it doesn't have a relative height of fifty feet and is linked to its parent peak by a very high col. Whitchurch Common covers most of its upper slopes; several Bronze Age stone hut circles are situated on its west-northwestern flank; a number of disused quarries are located on its western one and the sixteenth century Beckamoor Cross (or Windy Post) is sited at the bottom of its southern one.

210. (Westcott Hill), 1,098.75 feet/334.90 metres, SX 787863, a somewhat elongated, partly rocky peak, which rises steeply from a col to the east, quite steeply and then gently from a dry valley to the south, very gently from a col to the west and variably from a tributary dale of the River Teign to the north-

northwest. The hill is unnamed on the OS 1:25000 and 1:50000 maps, but it seems appropriate to name it after the hamlet of Westcott and Westcott Wood, which are both situated on its slopes. OS Maps shows its summit to be at the OS 1:25000 map trig point, which is marked as standing at 335 metres. Blackingstone Quarry is situated on its south-southwestern flank; there is a fair amount of woodland on its slopes and the B 3212 runs across its lower northern ones, broadly from west-southwest to east-southeast.

211=. [Ingra Tor], 1,093.50 feet/333.30 metres, SX 555721, a small, very rocky peak on a west-northwest-projecting spur of Leeden Tor. It rises fairly steeply and then steadily from the Walkham Valley to the west, steadily from one of its tributary dales to the north and very gently and then quite steeply from another to the south. OS Maps shows its highest point to be on the top of its summit tor, by the OS 1:25000 map 339-metre spot height. The fair discrepancy in the peak's altitude given by the online and paper versions of the OS's data might be explained by a limitation of the website's programme to calculate accurately the heights of rapidly-rising features. It doesn't have a relative height of fifty feet and is linked to its parent peak by a high col. The east of its summit area plunges into an abandoned quarry; a Bronze Age stone hut circle is situated on its west-southwestern flank and there is a fair amount of woodland on some of its lower slopes.

211=. Western Beacon, 1,093.50 feet/333.30 metres, SX 654577, an ovalish-shaped, pretty rocky peak, which is the most southerly 1,000-foot hill in Devon. It rises increasingly steeply from a col to the south, pretty steadily from the Lud Valley to the east, very gently from Black Pool at a col to the north and variably from the Erme Valley to the west. OS Maps shows its summit to be just south of the OS 1:25000 map 334-metre spot height. A number of Bronze Age cairns are situated on its upper southern flank; a stone row is located on its lower one and the Cuckoo Ball Neolithic chambered tomb is sited on its lower northeastern one. There are several disused quarries on its slopes; Ivybridge is situated on its lower southwestern flank and Bittaford is located at the foot of its east-southeastern one. The A 38, B 3213 and a railway line run across its lower southern slopes, broadly from west-southwest to east-northeast, and there is a fair amount of woodland on its lower western flank.

213. Corringdon Ball, 1,090.88 feet/332.50 metres, SX 670609, a round, slightly rocky peak, which rises fairly steadily from the Glaze Valley to the south, the East Glaze Valley to the west and the Badworthy Valley to the east and gently from a col to the north-northwest. OS Maps shows its summit to be approximately 35 yards south of the OS 1:25000 map 334-metre spot height. There are several pillow mounds and disused pits on its flanks; Treeland Downs and a prehistoric hut circle are situated on its north-northeastern one and a few small areas of woodland are located on its slopes.

214. Bratton Down, 1,089.90 feet/332.20 metres, SS 662392, an elongated, very small plateau, which rises fairly steeply and then very gently from a tributary valley of the River Bray to the east, quite gently and then very gently from another to the south, very gently from cols to the south-southwest and north and variably from a tributary dale of the River Yeo to the west. OS Maps shows its summit to be on its southern plateau, just north of the northern end of a reservoir, surprisingly approximately 175 yards west-northwest of the OS 1:25000 map 330-metre tiny contour ring. A telecommunications mast is situated on its summit area and the remains of a medieval village are found near the foot of its east-southeastern one. There are quite a number of small areas of woodland and one or two prehistoric remains on its slopes and the A 399 runs across the western side of its summit area, broadly from south to north.

215=. Manaton Rocks, 1,088.58 feet/331.80 metres, SX 748815, an elongated, rocky peak, which rises fairly steadily from a tributary dale of the River Bovey to the north, gently from a col to the west-northwest and variably from the Hayne Valley to the south and the Bovey Valley at Neadon Cleave to the east. OS Maps shows its summit to be at the OS 1:25000 map 330-metre spot height. There is a considerable amount of woodland on its slopes and the village of Manaton is situated on its south-southeastern flank.

215=. (Nattadon Hill), 1,088.58 feet/331.80 metres, SX 705866, a roundish, slightly rocky peak, which rises quite gently from a col to the east-southeast, variably from the Teign Valley to the north and one of its tributary dales to the south and quite steeply from the latter to the west. It is unnamed on the OS 1:25000 and 1:50000 maps, but it's often referred to as Nattadon Common (which covers most of its upper slopes) and the hamlet of Nattadon is situated on its northern flank, so Nattadon Hill seems to be an appropriate name for it. OS Maps shows its summit to be at the western edge of the OS 1:25000 map 333-metre spot height. The Iron Age Nattadon Common Fort is located on its summit area; there is a fair amount of woodland on its northern flank and Chagford is sited on its north-northwestern one. The B 3206 runs across its lower northern slopes, broadly from west-southwest to east-northeast.

217. [Hingston Down], 1,087.60 feet/331.50 metres, SX 770859, a small peak on a west-northwest-projecting spur of Pepperdon Down. It rises fairly steadily and then quite steeply from the head of a tributary valley of the Wray Brook to the north, fairly steadily from a dry dale to the east and variably from the Wray Valley to the west. OS Maps shows its summit to be in the middle of the OS 1:25000 map small 330-metre contour ring. However, it doesn't have a relative height of fifty feet and is linked to its parent peak by a high col. Hingston Rocks are situated on its west-southwestern flank and there are one or two small areas of woodland on its lower slopes.

218. [Smeardon Down],1,087.27 feet/331.40 metres SX 522781, a very small, pretty rocky peak on a west-projecting spur of White Tor, more than 1¼ miles to the east-northeast. It rises steadily from the Colly Valley to the south and variably from the Tavy Valley to the west and north. OS Maps shows its summit to be slightly north of the middle of the widest part of the OS 1:25000 map 330-metre contour bulge, just above the hill's name. However, the land drops by only 8.86

feet to a very high col linking it to its parent peak. Two prehistoric hut circles are situated on its slopes; Little Combe Tor is located on its south-southwestern flank; the remains of a prehistoric field system are sited on its western one; disused mine workings are found on its west-northwestern one and the hamlet of Cudliptown can be seen on its northern one. There are a number of patches of woodland on some of its lower slopes.

219. [Eastern Tor], 1,086.94 feet/331.30 metres, SX 584664, a somewhat elongated, fairly rocky peak on a south-southwest-projecting spur of Higher Hartor Tor, well over 1½ miles to the north-northeast. It rises fairly steadily from the Plym Valley to the south and east and quite gently from a depression to the west. OS Maps shows its summit to be at the OS 1:25000 map 333-metre spot height. However, it doesn't have a relative height of fifty feet and is linked to its parent peak by a high col. Ditsworthy Warren House is situated on its southern flank; several prehistoric settlements are located on its east-southeastern one; Gutter Mire lies at the foot of its north-northwestern one and disused tin workings are sited on its lower southwestern one.

220. [Vale Down], 1,084.97 feet/330.70 metres, SX 531859, a small, flat-topped peak on a south-projecting spur of Great Nodden, just over a mile to the north-northeast. It rises fairly steadily from the nearby Lyd Valley to the east, variably from the same dale to the south and gently from a tributary valley of the River Lew (via Battishill Down and the hamlet of Vale Down) to the west. OS Maps shows its summit to be at the OS 1:25000 map 331-metre spot height. However, it doesn't have a relative height of fifty feet and is linked to its parent peak by a high col. High Down is situated on its south-southwestern flank and Lydford Viaduct on its west-southwestern one. The A 386 runs across its western slopes, broadly from south to north.

221. Brent Tor, 1,083.33 feet/330.20 metres, SX 470804, an impressive, isolated, rocky, basalt peak, which is the most westerly 1,000-foot hill in Devon. It rises very gently and then steeply from a tributary dale of the River Lew to the west and variably from Burcombe to the north, the Burn Valley to the east and the head of one of its tributary dales to the south-southeast. OS Maps shows its summit, which is topped by St Michael's Church, to be at the OS 1:25000 map 334-metre spot height. The village of North Brentor is situated on its northeastern flank and there are a few small areas of woodland on its slopes.

222. Gidleigh Tor, 1,083.01 feet/330.10 metres, SX 671877, a roundish, rocky and heavily-wooded peak, topped by a semi-hidden tor. It rises steeply from the North Teign Valley to the south, variably from the same dale to the east, quite gently from a col to the west and very gently and then quite gently from the Moortown Valley (via the village of Gidleigh) to the north. OS Maps shows its summit to be on the western edge of the OS 1:25000 map 334-metre spot height. Gidleigh Park is situated on its southern and eastern flanks and Gidleigh Castle is located on its northern one.

223. [Yarde Down], 1,079.40 feet/329.00 metres, SS 722352, a small, flattish-topped peak on a broad, southwest-projecting spur of Five Barrows Hill, approaching 1¼ miles to the north-northwest. It rises quite steeply and then very gently from a tributary valley of the River Mole to the east, gently from another to the south and quite gently and then gently from one of the River Bray to the west. OS Maps shows its summit to be around a junction of four field boundaries, just north of the word "Down", as printed on the OS 1:25000 map within the 320-metre contour bulge. However, the land drops by only 5.25 feet to a very high col linking it to its parent peak. There is a fair amount of woodland on its lower slopes.

224. Hunters Tor, 1,067.59 feet/325.40 metres, SX 762823, an elongated, partly rocky, flattish-topped peak, which rises steeply and then very gently from the Bovey Valley at Lustleigh Cleave to the south, fairly gently and then steeply from the same dale to the west and variably from a tributary dale of the Wray Brook to the north and east. OS Maps shows its summit to be just south of the OS 1:25000 map 326-metre spot height, on the southeastern rampart of an Iron Age hillfort. The tor itself is located on the northeastern flank of the hill; Raven's Tor is located on its southern one; Harton Chest Rocks are sited on its south-southeastern one; Sharpitor and the Nut Crackers logan stone lie on its southeastern one and North Harton Down is found on its north-northeastern one. There is a considerable amount of woodland on its slopes, particularly to the south and south-southeast.

225. Grange Hill, 1,066.27 feet/325.00 metres, SS 661378, an elongated, flattish-topped peak, which rises increasingly gently from a tributary dale of the River Bray to the east and others of the River Yeo to the west (via the village of Bratton Fleming) and south and very gently and then quite gently from a col to the north-northwest. OS Maps shows its summit to be by a field boundary, approximately 200 yards southeast of Bratton Fleming Solar Farm and 125 yards south-southwest of the A 399 within the OS 1:25000 map 320-metre contour ring. The hill is unnamed on the OS 1:25000 and 1:50000 maps per se, but is evidenced by a minor road called Grange Hill, which cuts across its upper west-northwestern slopes. Several prehistoric barrows are situated on its summit area; the village of Benton is located on its lower south-southwestern flank and there are several small woods on its lower western one. The A 399 runs across its eastern and northern slopes, broadly from south-southeast to north-northwest.

226. Week Down, 1,063.98 feet/324.30 metres, SX 712863, an elongated, ovalish-topped peak, which rises gently from a col to the west-northwest and variably from the Bovey Valley to the south and a tributary dale of the River Teign to the north and east. OS Maps shows its summit to be approximately 50 yards south-southeast of the OS 1:25000 map 324-metre spot height. Week Down Cross, a historic way marker, is situated on its eastern flank; the hamlet of Middlecott is located on its southeastern one and there are several small woods on its lower slopes.

227. [Kingsett Down], 1,063.32 feet/324.10 metres, SX 521814, a small, flat-topped peak on a south to south-southwest-curving spur of Black Down, to the north. It rises increasing gently from the Tavy Valley (via the village of Horndon) to the south and the Cholwell Valley to the west and very gently from the Willsworthy Valley to the east. OS Maps shows its summit to be roughly in the middle of the 320-metre contour bulge, approximately at the top end of the letters "J" and "e", in the word "Jewell", as printed on the OS 1:25000 map. However, the land drops by only 9.84 feet to a very high col linking it to its parent peak. Wheal Jewell Reservoir is situated on the eastern side of its summit area; there are quite a number of disused shafts on its slopes and lines of woodland on some of its lower ones; Brimhill Tor and Kent's Tor are located on its south-southwestern flank and Horndon Down is sited on its east-southeastern one.

228. [Berry Hill], 1,061.35 feet/323.50 metres, SS 669373, a roundish peak, which rises steeply and then very gently from the Bray Valley to the east, quite steeply and then gently from one of its tributary dales to the south and quite steeply and then very gently from another to the north. OS Maps shows its summit to be just east-northeast of the 324-metre OS 1:25000 map spot height. However, it fails to be an independent hill by just 2.76 feet and instead is an outlier of nearby Grange Hill, to the west-northwest, to which it's linked by a high col. A number of prehistoric remains are situated on its greater summit area; its lower eastern and east-southeastern flanks are heavily wooded and Stock Down is located on the latter.

229. (Corndon Hill), 1,057.09 feet/322.20 metres, SX 690854, a wedge-shaped, roundish-topped peak, which rises fairly steadily from the head of a tributary dale of the South Teign River to the north and a col to the west, gently from another to the northeast and the head of a tributary dale of the West Bovey Valley to the east and variably from the latter to the south. The hill is unnamed on the OS 1:25000 and 1:50000 maps, but the small settlements of Higher Corndon, Lower Corndon and Western Corndon lie on its flanks, so Corndon Hill seems an appropriate name for it. OS Maps shows its summit to be approximately in the middle of the OS 1:25000 map 320-metre contour ring.

230. Barna Barrow, 1,056.76 feet/322.10 metres, SS 754495, an ovalish-shaped peak, which is the most northerly 1,000-foot hill in Devon. It rises steeply and then quite gently from Coddow Combe to the north and a tributary dale of the East Lyn Valley to the south, fairly steadily and then quite gently from a dry valley to the west, quite gently from a col to the east-southeast and gently from another to the west-northwest. OS Maps shows its summit to be just northwest of the OS 1:25000 map 323-metre spot height. The A 39 bends across its southern flank, broadly from east-southeast to west-northwest, and Holden Head is situated below it.

231. (Pinmoor Hill), 1,054.79 feet/321.50 metres, SX 756887, an elongated, partly rocky peak, which rises steeply and then quite gently from the Teign Valley to the north, quite steeply from the head of one of its tributary dales to the west and variably from another to the east and the Wray Valley to the south. The hill is unnamed on the OS 1:25000 and 1:50000 maps, but it seems appropriate to name it after Pinmoor Farm, which is situated on its eastern flank. OS Maps shows its summit to be just north of the OS 1:25000 map 323-metre spot height. Its lower northern slopes are heavily wooded and there are smaller areas of woodland on some of its other ones. Wooston Castle Iron Age hillfort is situated on its northeastern flank.

232. Trentishoe Down, 1,054.13 feet/321.30 metres, SS 629477, a somewhat elongated, roundish-topped peak, which rises very steeply and then quite steeply from the Bristol Channel (via the cliffs of Elwill Bay) to the north, gently from a col to the west-southwest, fairly steadily from a tributary valley of the River Heddon to the south and variably from another to the east. OS Maps shows its summit to be on the top of the more northerly of two Bronze Age barrows, at the OS 1:25000 map 323-metre spot height. The village of Trentishoe is situated on its east-northeastern flank; North Cleave and North Cleave Gut are located on its northeastern one and there are several (mainly small) areas of woodland on its lower slopes.

233. [Caffyns Heanton Down], 1,049.87 feet/320.00 metres, SS 686468, an elongated, flattish-topped peak, which rises quite steeply and then gently from Croscombe to the north, fairly steadily and then gently from Ranscombe to the east, increasingly gently from a dry valley to the east-northeast, gently from a col to the west-northwest and very gently from another to the south-southwest. OS Maps shows its summit to be approximately 50 yards southeast of the OS 1:25000 map 320-metre spot height. However, it fails to be an independent hill by only 8.99 feet and instead is an outlier of Butter Hill, approaching two miles to the south-southeast, to which it's linked by high col. Wildner Top is situated on its east-northeastern flank; a line of woodland covers most of its lowest northern and north-northeastern slopes and the A 39 runs across its southern and eastern ones, broadly from west-southwest to east-northeast.

234. Pew Tor, 1,049.54 feet/319.90 metres, SX 532734, an impressive, very rocky peak, which rises gently and then fairly steadily from a tributary dale of the River Tavy to the west, gently from a col to the north-northeast and variably from the Walkham Valley to the east and one of its tributary dales to the south. OS Maps shows its highest point to be on the top of its summit tor, within the largest of three OS 1:25000 map 320-metre very small contour rings. One or two disused quarries are situated on its slopes; there is a considerable amount of woodland on its lower eastern and southeastern ones and the hamlet of Sampford Spiney is located at the foot of its southern flank.

235. (Plaston Green Hill), 1,048.56 feet/319.60 metres, SX 795868, a short, south-southeast to east-northeast, clockwise-running, flat-topped ridge, which rises fairly steeply and then gently from the village of Westcott and a tributary valley of the River Teign to the west, fairly steadily from another such dale to the north and a col to the east and variably from another col to the south-southeast. The hill is unnamed on the OS 1:25000 and 1:50000 maps, but it seems appropriate to name it after Plaston Green, the nearest OS map place name, which is on the peak's eastern flank. OS Maps shows the hill's summit to be approximately 50 yards west-southwest of the OS 1:25000 map 320-metre spot height. A minor road cuts across its summit area and a number of small woods are situated on its lower slopes.

236. [High Down], 1,045.93 feet/318.80 metres, SX 530855, a south-projecting spur of Great Nodden (and, more immediately, Vale Down), more than 1¼ miles to the north-northeast. It rises quite steeply and then very gently from the Lyd Valley to the south, fairly gently from the same dale to the east and gently from a tributary valley of the River Lew to the west. OS Maps shows its summit to be at the OS 1:25000 map 320-metre spot height. However, the land drops by only 11.15 feet to a very high col linking it to its parent peak. The A 386 runs across its western slopes, broadly from south-southwest to north-northeast.

237. [Bench Tor], 1,045.60 feet/318.70 metres, SX 691716, a short, south to north-running, pretty rocky escarpment on a north-projecting spur of Ryder's Hill, 2½ miles to the west-southwest. It rises very steeply from the Dart Valley to the east, very steeply and then gently from the same dale to the north and pretty steeply and then fairly steadily from the Venford Valley to the west. OS Maps shows its highest point to be on the top of its summit tor, towards the southeastern end of the OS 1:25000 map 320-metre very small contour ring. However, it doesn't have a relative height of fifty feet and is linked to its parent peak by a high col. The remains of ancient homesteads are situated on its southwestern flank, below which is Venford Reservoir; the Dart Valley Nature Reserve is located on its northern one and Sharrah Pool lies at the foot of its eastern one. Most of its lower slopes are covered by woodland.

238. Ausewell Rocks, 1,045.28 feet/318.60 metres, SX 735717, a wedge-shaped, partly rocky peak, which rises very steeply and then quite steeply from the Dart Valley to the west, variably from the same dale (via Ausewell Common) to the south and the Ashburn Valley to the east and very gently from a col to the north-northeast. OS Maps shows its highest point to be on the top of its summit tor, at the OS 1:25000 map 325-metre spot height. The considerable discrepancy in the peak's altitude given by the online and paper versions of the OS's data might be explained by a limitation of the website's programme to calculate accurately the heights of rapidly-rising features. There are large areas of woodland and several disused shafts on its slopes; a number of Bronze Age cairns are situated on its summit area; Cleft Rock is located on its south-southwestern flank; Raven Rock is sited on its west-southwestern one; Lover's Leap is found on its west-northwestern one and Boro' Wood Castle Iron Age enclosure can be seen on its lower eastern one.

239. Laployd Hill, 1,043.64 feet/318.10 metres, SX 806850, an irregular-shaped peak, which rises fairly steeply and then fairly steadily from a tributary dale of Kennick Reservoir to the south and variably from a dry dale to the east, the Rookery Valley to the north and the Kennick Valley to the west. The hill is unnamed on the OS 1:25000 and 1:50000 maps, but it has recently become named after

Laployd Plantation, which covers its summit area, nearly all its southern and western slopes and its upper eastern ones. OS Maps shows its (wooded) highest point to be on the eastern edge of the OS 1:25000 map 320-metre small contour ring. Beacon Plantation is situated on its eastern and east-northeastern flanks and Kennick Reservoir at the foot of its southwestern and south-southwestern ones.

240. [Heckwood Tor], 1,042.98 feet/317.90 metres, SX 537738, an ovalish-shaped, pretty rocky peak, which rises variably from the Walkham Valley to the east, gently from one of its tributary dales to the south and fairly gently from one of the River Tavy to the west. OS Maps shows its highest point to be on the top of its summit tor, at the OS 1:25000 map 321-metre spot height. However, it fails to be an independent hill by just 3.08 feet and instead is an outlier of Cox Tor (and, more immediately, Barn Hill and Feather Tor) more than 1½ miles to the north-northwest. Several disused quarries are situated on its slopes and there is a considerable amount of woodland on its lower ones.

241. Lynton Hill, 1,040.68 feet/317.20 metres, SS 705485, a roundish, slightly rocky peak, which rises very steeply and then increasingly gently from the West Lyn Valley to the east and south, increasingly gently from an unnamed dale to the west and variably from the Bristol Channel and Wringcliff Bay (via Wringcliff and The Valley of Rocks) to the north. The hill is unnamed on the OS 1:25000 and 1:50000 maps, but it has recently become named after the town of Lynton, which lies at the foot of its east-northeastern flank. OS Maps shows its highest point to be at the 318-metre OS 1:25000 map spot height. A trig point, a prehistoric cairn and a telecommunications mast are all situated on its summit area, which is crossed by Lydiate Lane; an early medieval inscribed stone is located on its west-southwestern flank; Duty Point Tower and Duty Point are sited on its northwestern one; South Cleave is found on its north-northwestern one; Hollerday Hill can be seen on its north-northeastern one and there is a fair amount of woodland on its lower slopes. The A 39 runs along the foot of its southern slopes, broadly from west-southwest to east-northeast, and the B 3234 crosses its lower eastern ones, essentially from south-southwest to north-northeast.

242. Great Hangman (Girt Down), 1,040.35 feet/317.10 metres, SS 600480, a wedge-shaped, round-topped peak, which rises very steeply and then steeply from the Bristol Channel and Blackstone Bay (via the highest sea cliff in England) to the north, very steeply and then fairly steeply from Sherrycombe to the east and variably from a col to the west and a tributary valley of the River Umber to the south. OS Maps shows its summit to be at the OS 1:25000 map cairn, which is a mere 370 yards from the sea and marked as standing at 318 metres. A prehistoric standing stone is situated on its southern flank; Knap Down is located on its south-southwestern one; the village of Combe Martin is sited on its southwestern one; The Rawn's cliffs and Rawn's Rocks are found on its west-northwestern one; Blackstone Beach lies at the foot of its northwestern one and Blackstone Point can be seen at the end of its north-northeastern one. There are quite a number of small areas of woodland on its lower inland slopes.

243. (Broadaford Hill), 1,032.81 feet/314.80 metres, SX 691767, a somewhat elongated, flattish-topped peak, which rises quite gently and then gently from the West Webburn Valley to the west, quite gently and then very gently from the same dale to the south, quite gently from the Broadaford Valley to the east and very gently from a col to the north. The hill is unnamed on the OS 1:25000 and 1:50000 maps, but it seems appropriate to name it after Broadaford, the nearest settlement, on its southern flank. OS Maps shows the hill's summit to be just east-northeast of the OS 1:25000 map 315-metre spot height.

244. [Beetor], 1,029.53 feet/313.80 metres, SX 708843, a small, round peak at the northern end of a north-projecting spur of Hookney Tor (and, more immediately, Shapley Tor), approaching two miles to the south-southwest. It rises variably from the Bovey Valley to the east, fairly steeply and then fairly gently from the same dale to the north-northwest and very gently and then quite gently from one of its tributary valleys to the west. OS Maps shows its summit to be at the OS 1:25000 map 317-metre spot height, but it doesn't have a relative height of fifty feet and is linked to its parent peak by a high col. The hill is unnamed on the OS 1:25000 and 1:50000 maps, but the proximity of Beetor Farm, Beetor Cross and Beetor Bridge make it clear that this peak is indeed Beetor. There are several small woods on its lower slopes and the B 3212 runs across its eastern ones, broadly from west-southwest to east-northeast.

245. Burnicombe Down, 1,028.54 feet/313.50 metres, SX 803869, an oval-shaped, partly rocky peak, which rises fairly gently from a col to the west, quite gently and then very gently from another to the east and variably from the Teign Valley to the north and the Rookery Valley to the south. OS Maps shows its summit to be at the OS 1:25000 map 315-metre spot height, on the top of a rocky outcrop. Bridford Wood is situated on its lower northern flank; there is a narrow line of woodland along its lowest southern slopes and the B 3212 runs across its lower northern ones, broadly from east to west.

246. [Alse Barrow], 1,027.56 feet/313.20 metres, SS 749446, a knoll on a short, north-northwest-projecting spur of Exe Plain, 1½ miles to the south. It rises steeply from the Farley Valley to the east and north, quite gently from the same dale to the northwest and quite steeply and then gently from one of its tributary valleys to the west. OS Maps shows its summit to be at the eastern side of the OS 1:25000 map 310-metre very small contour ring. However, it doesn't have a relative height of fifty feet and is linked to its parent peak by a high col.

247. (Howton Hill), 1,024.93 feet/312.40 metres, SX 751872, a somewhat elongated, roundish--topped peak, which rises quite steeply and then quite gently from the Wray Valley to the east, fairly steadily and then quite gently from a tributary dale of the River Teign to the west, very gently from a col to the north-northeast and variably from the Wadley Valley and Moretonhampstead to the south. The hill is unnamed on the OS 1:25000 and 1:50000 maps, but it seems appropriate to name it after the nearby hamlet of Little Howton to the west. OS Maps shows its summit to be at the OS 1:25000 map 312-metre spot height. There is a fair amount of woodland on its slopes and the A 382 and the B 3212 run across its lower southern ones, the former broadly from east-southeast to west-northwest and the latter essentially from east to west.

248=. [Feather Tor], 1,022.31 feet/311.60 metres, SX 534741, a small, very rocky peak on a south-southeast-projecting spur of Cox Tor (and, more immediately, Barn Hill), well over 1¼ miles to the north-northwest. It rises fairly gently from a tributary valley of the River Walkham to the east, very gently and then quite gently from one of the River Tavy to the west and very gently from a col to the south. OS Maps shows its highest point to be on the top of its summit rocks, a little north of the OS 1:25000 map 313-metre spot height. However, it doesn't have a relative height of fifty feet and is linked to its parent peak by a very high col. There are one or two small areas of woodland on its western slopes.

248=. Mockham Down, 1,022.31 feet/311.60 metres, SS 667357, an oval-shaped peak, which rises increasingly gently from a tributary valley of the River Bray to the east, very gently and then quite gently from the same dale to the north and very gently and then gently from cols to the west and south. OS Maps shows its highest points to be within the OS 1:25000 map 310-metre contour ring, on the top of the earthwork on both sides of the eastern gate of an Iron Age hillfort. There are small woods just to the west and north of its summit area and the A 399 runs across its northern, eastern and east-southeastern slopes, broadly from southeast to northwest.

250. Heltor Rock, 1,020.01 feet/310.90 metres, SX 799870, a small, round, partly rocky peak, capped by an impressive split tor. It rises increasingly steeply from the Rookery Valley to the south and a col to the west-southwest, fairly steadily from another to the east and variably from the Teign Valley to the north. OS Maps shows its summit to be on the top of the tor, towards the eastern side of the OS 1:25000 map 310-metre very small contour ring. Its lower northern and southern slopes are heavily wooded and the B 3212 runs across its lower northern flank, broadly from east to west.

251. (Barton Hill), 1,018.04 feet/310.30 metres, SS 683409, an elongated, flattish-topped peak, which rises fairly steadily and then gently from the Bray Valley to the south, quite steeply and then very gently from its tributary dales to the east and west and gently from a col to the north-northeast. The hill is unnamed on the OS 1:25000 and 1:5000 maps, but it seems appropriate to name it after the several "Barton" place and feature names situated around it. OS Maps shows its summit to be at the OS 1:25000 map 312-metre spot height. There is an area of woodland on its southeastern flank and the B 3358 runs across its northern one, broadly from east-southeast to west-northwest.

252. Cherryford Hill, 1,012.47 feet/308.60 metres, SS 674473, an elongated, round-topped peak, which rises increasingly gently from a tributary valley of the River Heddon to the south, quite gently from a col to the west, very gently from Croscombe to the east and variably from the Bristol Channel and

Woody Bay to the north. OS Maps shows its summit to be at the OS 1:25000 map trig point, which is marked as standing at a height of 309 metres. Martinhoe Common covers most of its upper slopes; the hamlet of Kemacott is situated on its west-southwestern flank; that of Martinhoe, The Beacon Roman fortlet, Highveer Point, Great Burland Rocks and The Cow and Calf rocks are all located on its north-northwestern one; the Wringapeak headland is sited on its lower northern one; Crock Point can be seen at the end of its north-northeastern one and Bonhill Top is found on its northeastern one, with Lee Bay below it. There is a fair amount of woodland on its lower northern and east-northeastern slopes.

253. Brent Hill, 1,007.87 feet/307.20 metres, SX 702616, an isolated, elongated, partly rocky peak, which rises increasingly gently from a tributary valley of the River Avon to the west, very gently and then steeply from a col to the north and variably from South Brent to the south and the Horse Valley to the east. OS Maps shows its highest point to be on the top of its summit rocks, just to the south-southwest of the OS 1:25000 map trig point, which is marked as standing at 311 metres. The remains of an Iron Age hillfort, with a defensive embankment, are situated on and around its summit area and Beara Common covers most of its upper southern slopes. A railway line, the B 3372 and the A 38 run across its southern flank, broadly from west-southwest to east-northeast.

254. (Myrtleberry Hill), 1,004.59 feet/306.20 metres, SS 739482, an oval-shaped, flattish-topped peak, which rises very steeply and then gently from the East Lyn Valley to the north, steeply and then fairly steadily from Hoar Oak Water to the east, steeply and then gently from the latter to the south, quite gently from a dry dale to the west and very gently from a col to the southwest. The hill is unnamed on the OS 1:25000 and 1:50000 maps, but it seems appropriate to name it after the Myrtleberry Cleave gorge, which is situated at the foot of its northern slopes. OS Maps shows its summit to be at the OS 1:25000 map 308-metre spot height. The remains of two Iron Age settlements are located on its eastern and north-northeastern flanks; its lower eastern and northern ones are heavily wooded and Watersmeet is sited at the foot of its northeastern one. The A 39 bends round its lower eastern and northern slopes, broadly from southeast to northwest.

255. [Legis Tor], 994.09 feet/303.00 metres, SX 571655, a small, rocky peak on a south-southwest to west-southwest-curving spur of Ringmoor Down, to the north-northeast. It rises fairly steadily and then very gently from the Plym Valley to the east and fairly gently from the same dale to the south and Legis Lake to the west. OS Maps shows its highest point to be on the top of its summit tor, at the OS 1:25000 map 310-metre spot height, but the land drops by only 9.18 feet to a very high col linking it to its parent peak. However, it doesn't qualify as a 1,000-foot hill according to OS Maps, which computes its height as being 7 metres lower than that marked on the map. This discrepancy might be explained by a limitation of the website's programme to calculate accurately the heights of rapidly-rising features. Legistor Warren and a series of pillow mounds are situated on its southern flank; two prehistoric cairns are located on its eastern one and disused tin workings are sited on its west-southwestern one.

256. [Vixen Tor], 985.24 feet/300.30 metres, SX 541742, a small, elongated, very rocky peak on a south-projecting spur of Great Staple Tor (and, more immediately, Middle Staple Tor), more than a mile to the north. It rises variably from the Walkham Valley to the east, gently from one of its tributary dales to the west and fairly steadily from the latter to the south. OS Maps shows its highest point to be on the top of its summit tor, a little north-northwest of the middle of the OS 1:25000 map 310-metre contour ring. However, it doesn't have a relative height of fifty feet and is linked to its parent peak by a very high col. Also, it doesn't qualify as a 1,000-foot hill according to OS Maps, which computes its height as being 16.70 metres lower than that marked on the map! This discrepancy might be explained by a limitation of the website's programme to calculate accurately the heights of rapidly-rising features. A Bronze Age cist is situated on its northern flank and a narrow strip of woodland runs along most of its lowest slopes.

5 Somerset's 1,000-Foot Peaks

Somerset is a very varied county geographically, with large expanses of flat land (especially the Somerset Levels), mainly in the centre of the shire, sandwiched between upland areas to the west and southwest and the north. The largest hilly area is Exmoor, in the far west, part of which spills over into Devon, and it contains well over half the county's 1,000-foot peaks, including its highest ones. The Brendon Hills and one or two outliers and then the Quantock Hills, which are all likewise composed of sedimentary rocks, are situated to the east of Exmoor and most of the rest of Somerset's 1,000-foot hills are found on their ranges. Finally, a handful of the county's 1,000-foot peaks are situated on the Carboniferous limestone Mendip Hills, to the north of the Somerset Plain, and the sedimentary Blackdown Hills, to its southwest, which straddle the border with Devon.

1. Dunkery Hill (Dunkery Beacon), 1,698.49 feet/517.70 metres, SS 891415, a very large, somewhat elongated, oval-topped peak, which is the highest hill on Exmoor and in Somerset. It rises pretty steeply from the East Valley to the north, quite steeply and then fairly steadily from the Spangate Valley to the east, fairly steadily from the Quarme Valley to the south and gently from a col to the west-southwest. OS Maps shows its highest point to be on the summit of the Bronze Age Dunkery Beacon cairn (which itself is topped by a modern cairn), nearby a trig point, which is marked on the OS 1:25000 map as standing at 519 metres. The remains of a Bronze Age barrow cemetery are situated on its summit area; Kit Barrows Bronze Age burial cairns are located on its east-northeastern flank and two Iron Age enclosures and a deserted medieval settlement are sited on its northern one. Luccombe Hill is found on its east-northeastern flank; the source of the River Avill can be seen on its southern one and there is a fair amount of woodland on its lowest slopes.

2. Honeycombe Hill (Great Rowbarrow), 1,672.90 feet/509.90 metres, SS 875414, a large, elongated peak, with numerous spurs projecting in different directions. It rises steadily from the Quarme Valley to the south, very gently from a col to the east and variably from another to the west and Lang Combe to the north. OS Maps shows its summit to be approximately just north of the digit "0", as printed in the OS 1:25000 map unmarked 510-metre spot height number. Numerous prehistoric remains are situated on its summit area and flanks and include Great and Little Rowbarrow Bronze Age burial cairns, Bendels (Bronze Age) Barrows and Bagley Iron Age enclosure. Codsend Moors cover most of its southern slopes; Hoar Moor, the source of the River Quarme and Exford Common are located on its west-southwestern flank; Wilmersham Common is sited on its west-northwestern one; Stoke Pero Common and Stoke Ridge are found on its northern one and Goosemoor Common is positioned on its northeastern one. There is a considerable amount of woodland on the lower slopes of its northern hemisphere.

3. The Chains, 1,595.47 feet/486.30 metres, SS 734419, an elongated plateau, which rises quite steeply and then very gently from the Warcombe Valley to the north and the Hoaroak Valley to the north-northeast, increasingly gently from

The Chains Valley to the east, gently and then very gently from the Barle Valley to the west and variably from one of the tributary dales of the last named to the south. OS Maps shows its summit to be on the top of the Bronze Age Chains (bowl) Barrow, by the OS 1:25000 map trig point, which is marked as standing at 487 metres. The hill is a major watershed, with the source of the River Barle on its west-northwestern flank, that of the West Lyn River on its northwestern one and Exe Head in a small col just below its east-southeastern one. Quite a number of prehistoric remains, including standing stones, stone rows and cairns, are situated on its slopes. Goat Hill is located on its southwestern flank; Pinkworthy is sited on its western one; Pinkery Pond lies at the foot of its west-northwestern one; Thorn Hill is found on its north-northwestern one and Hoaroak Hill and Benjamy can be seen on its north-northeastern one. The B 3358 runs across its lowest southern slopes, broadly from east-southeast to west-northwest.

4. Hangley Cleave, 1,581.04 feet/481.90 metres, SS 747362, an elongated, flattish-topped peak, which rises steeply and then very gently from the Kinsford Valley (via Long Holcombe) to the east, steeply and then gently from the same dale to the north, fairly steadily and then quite gently from a tributary valley of the River Mole to the south and very gently from a col to the west-northwest. OS Maps shows its highest point to be approximately in the middle of the OS 1:25000 map 480-metre contour ring. The county boundary with Devon runs across the western side of its summit area, broadly from east-southeast to west-northwest. Four prehistoric bowl barrows are situated on its summit area, the largest two of which are known as Two Barrows. Fyldon Common is located on its west-southwestern flank; there is a fair amount of woodland on its lowest southern and south-southeastern slopes and Span Head is sited at the foot of its west-southwestern one.

5. Dure Down, 1,571.19 feet/478.90 metres, SS 758413, an oval-shaped peak, which rises steeply and then very gently from the Exe Valley to the east, quite steeply and then gently from the same dale to the north, very gently from Exe Head to the west-northwest and variably from the Bale Valley to the south. OS Maps shows its summit to be approximately 60 yards east-southeast of the OS 1:25000 map 480-metre spot height. The village of Simonsbath is situated at the foot of its south-southeastern flank; the B 3358 runs across its lowest southern slopes, broadly from east to west, and the B 3223 crosses its southeastern and eastern ones, essentially from south-southeast to north-northwest.

6. [Ricksy Ball (Squallacombe Head)], 1,552.82 feet/473.30 metres, SS 726385, a small plateau, which rises fairly steeply and then gently from Mel Combe (via Great Melcombe) and Henthitchen Combe to the west and Squallacombe to the east and fairly steadily and then gently from Great Vintcombe to the north. OS Maps shows its summit to be approximately in the middle of the northern part of the OS 1:25000 map 470-metre large contour ring. However, it fails by only 5.05 feet to be an independent peak and instead is an outlier of Five Barrows Hill, approaching 1¼ miles to the south-southeast, to which it's linked by a high col. The county boundary with Devon runs across the southern end of its summit area and its upper western slopes, broadly from south-southeast to north-northwest. The Bronze Age Setta (bowl) Barrow is part of a cairn cemetery, which is situated at the southern end of Ricksy Ball's summit area, and a series of prehistoric standing stones are located on its lower east-southeastern flank. Muxworthy Ridge and Bray Common are sited on its west-southwestern flank and Black Hill is found on its northwestern one.

7. Great Hill, 1,526.25 feet/465.20 metres, SS 844429, a narrow, elongated plateau, which rises increasingly gently from the head of the Weir Valley to the north, pretty gently from the Chetsford Valley to the south, very gently from Madacombe to the west and a col to the west-southwest and variably from Hawk Combe to the north-northeast and the Nutscale Valley to the east. OS Maps shows its summit to be just south of the digits "6" and "2", as printed in the OS 1:25000 map 462-metre spot height. There are a number of prehistoric remains on its slopes, including Berry Castle, an Iron Age or Romano-British hillfort on its lower north-northeastern flank. Alderman's Barrow Allotment is situated on its southern flank; Little Hill is located on its east-southeastern one; Lucott Moor and Babe Hill are sited on its east-northeastern one, at the foot of which is Nutscale Reservoir, and Black Mires lie on its upper western one.

8. Little Ashcombe, 1,498.69 feet/456.80 metres, SS 782406, an elongated, flat-topped peak, which rises steeply and then gently from the Exe Valley to the east and north, very gently from a col (via Prayway Head and Great Ashcombe [sic]) to the west and variably from the Barle Valley to the south. OS Maps shows its summit to be at the OS 1:25000 map 457-metre spot height. Three Combe Hill and Exe Cleave are situated on its east-southeastern flank; Raven's Nest rock face is located on its west-northwestern one and there are a number of small areas of woodland on its lower slopes.

9. Exe Plain, 1,493.77 feet/455.30 metres, SS 754422, a small plateau, which rises quite steeply and then gently from the Hoaroak Valley to the west, gently from the Exe Valley to the south and very gently from cols to the north-northwest and east. OS Maps shows its summit to be at the OS 1:25000 map 456-metre spot height. Quite a number of prehistoric remains, including standing stones and stone rows, are situated on its northern flank and the Hoar Oak Tree, an ancient boundary marker, is located on its lower northwestern one.

10. [Hurdle Down], 1,487.86 feet/453.50 metres, SS 841419, an elongated, flat-topped peak, which rises quite steeply and then gently from the Chetsford Valley to the east, increasingly gently from the same dale to the north, fairly steadily and then gently from the Greenland Valley to the south and quite gently and then gently from the head of the Allcombe Valley to the west. OS Maps shows its summit to be approximately 75 yards southwest of the OS 1:25000 map 453-metre spot height. However, it fails by just 2.76 feet to be an independent hill and instead is an outlier of Great Hill, to the north, to which it's linked by a high col. There are a number of prehistoric remains on its slopes, including stone alignments on its south-southeastern flank and the Bronze Age Alderman's (round)

Barrow on its west-northwestern one. Almsworthy Common is situated on its upper slopes; Wellshead Allotment and an area of woodland are situated on its west-southwestern flank and several disused quarries are located on its slopes.

11. [Hoar Tor (Little Buscombe)], 1,470.80 feet/448.30 metres, SS 767424, a small, elongated plateau, which rises fairly steeply and then very gently from Lanacombe to the east, quite gently and then very gently from the Exe Valley to the south, quite gently and then gently from a tributary dale of Hoccombe Water to the north and gently and then very gently from a col to the north-northwest. OS Maps shows its summit to be at the OS 1:25000 449-metre spot height. However, it fails to be an independent hill by just 5.05 feet and instead is an outlier of Exe Plain, to the west, to which it's linked by a high col. The remains of Rexy Barrow and another Bronze Age bowl barrow are situated on the southern part of its summit area and Great Buscombe is located on its east-southeastern flank. The county boundary with Devon crosses its northern slopes, broadly from west-southwest to east-northeast, and the B 3223 cuts across its western one, essentially from south to north.

12. Swap Hill (Elsworthy), 1,455.71 feet/443.70 metres, SS 812415, a flat-topped peak at the summit of a narrow, elongated plateau. It rises increasingly gently from Rams Combe to the south and the Sparcombe Valley to the south-southeast, gently from a tributary dale of Long Combe to the north and very gently from cols to the west-southwest and east-northeast. OS Maps shows its summit to be at the OS 1:25000 map trig point, which is marked as standing at 444 metres. Dry Hill is situated to the west-southwest; there are several disused quarries on its slopes and the remains of prehistoric stone rows are located at two sites on its north-northwestern flank.

13. [Dry Hill (Ware Ball)], 1,454.72 feet/443.40 metres, SS 804411, a flat-topped peak, which shares an elongated plateau with the higher Swap Hill, to the east-northeast, to which it's linked by a very high col. It rises steeply and then very gently from the Exe Valley to the south, variably from the same dale to the west and pretty gently from the Badgworthy Valley to the north. OS Maps shows its summit to be just to the southwest of the OS 1:25000 map 444-metre spot height. However, it doesn't have a relative height of fifty feet and is linked to its parent peak by a very high col. A prehistoric cairn lies at its summit; a long barrow is situated on its lower southern flank and various antiquities are present on its north-northeastern one. East Pinford is located on its north-northwestern flank, West Pinford on its west-northwestern one and The Warren on its lower western one.

14. Horsen Hill, 1,452.76 feet/442.80 metres, SS 789356, an elongated, flattish-topped peak, which rises variably from the Sherdon Valley to the south, steeply and then very gently from the same dale to the east, steeply and then gently from the Kinsford Valley to the west, fairly steadily from a tributary dale of the River Barle to the north and gently from a col to the north-northwest. OS Maps shows its highest point to be on the top of its summit Bronze Age round barrow, just to the north of the OS 1:25000 map 443-metre spot height. Its summit cairn is one of Sherdon Barrows, a group of prehistoric remains on the hill. Sherdon is situated on its east-northeastern flank and Great Ferny Ball is located on its north-northeastern one.

15. Porlock Hill, 1,430.77 feet/436.10 metres, SS 848459, a somewhat elongated, flattish-topped peak, which rises fairly steeply and then quite gently from Pitt Combe to the north, Shillett Combe to the east and Hawk Combe to the south and very gently from a col to the west-southwest. The peak is most commonly known for the remarkable 1 in 4 climb, with hairpin bends, of the A 39 from Porlock up its eastern flank (before the road cuts across its northern one). OS Maps shows its summit to be at the western edge of the OS 1:25000 map 436-metre spot height. Porlock Common is situated on its upper slopes; the Whit Stones, a pair of possibly prehistoric standing stones, are located on its upper east-northeastern flank; Porlock is sited at the foot of its east-northeastern one; Porlock Bay is found at the bottom of its northeastern one and there is a considerable amount of woodland on its lower slopes.

16. [Long Holcombe], 1,430.45 feet/436.00 metres, SS 771359, a small, flat-topped peak on an east-projecting spur of Hangley Cleave, approaching 1½ miles to the west. It rises steeply and then gently from the Kinsford Valley to the north, pretty steeply and then very gently from the same dale to the east and quite gently from a tributary valley of Sherdon Water to the south. OS Maps shows its summit to be approximately in the middle of the widest part of the OS 1:25000 map 430-metre contour bulge, but the land drops by only 9.52 feet to a very high col linking it to its parent peak.

17. Culbone Hill, 1,421.26 feet/433.20 metres, SS 837466, a large, elongated peak, which rises steeply and then fairly steadily from Smalla Combe to the north, steeply and then quite gently from Pitt Combe to the east, steeply and then gently from Met Combe to the west and fairly steadily and then gently from Ven Combe to the south. OS Maps shows its summit to be at the OS 1:25000 map trig point, which is marked as standing at 433 metres. A telecommunications mast is situated on its greater summit area and large parts of the hill are wooded. A series of prehistoric remains are located on its upper northwestern and north-northwestern flanks and include a stone row and the Bronze Age Quarter Barrow. The early medieval inscribed Culbone Stone is sited in the same area. The A 39 runs across its upper southern and western slopes, broadly from east-southeast to west-northwest.

18=. (Brightworthy Hill), 1,401.25 feet/427.10 metres, SS 818350, a somewhat elongated, roundish-topped peak, which rises fairly steadily and then quite gently from the Barle Valley to the north, increasingly gently from Dillacombe to the west, gently and then very gently from a tributary dale of Litton Water (via Halscombe Allotment and Hawkridge Common) to the south and variably from Knighton Combe to the east. OS Maps shows its highest point to be at the OS 1:25000 map trig point, which is marked as standing at 428 metres. It is situated on the top of one of the two (of three) surviving Bronze Age Brightworthy Barrows located on its summit area and the remains of Green (bowl) Barrow lie on its upper southern flank. The peak is unnamed on the OS 1:25000 and 1:50000 maps, but is sometimes called Withypool Common (which is sited on its east-southeastern flank), although this is the name of the land and not the hill. As a Withypool Hill already exists, it seems more appropriate for the peak to be named after its defining summit features, its barrows. Black Pits Plain is situated on its south-southwestern flank and Worth Hill on its southeastern one.

18=. Winsford Hill, 1,401.25 feet/427.10 metres, SS 877342, an elongated, round-topped peak, which rises very steeply and then gently from the Punchbowl to the east, fairly steadily and then quite gently from the Winn Valley to the north, quite gently from the Little Valley (via the hamlet of Liscombe) to the south and variably from a tributary dale of the River Barle to the west. OS Maps shows its summit to be just east of the OS 1:25000 map trig point, which is marked as standing at 426 metres. Five Bronze Age round barrows are situated on its summit area and the most westerly three are named The Wambarrows. The B 3223 runs across the south of its summit area, broadly from east-southeast to west-northwest; parts of its lower slopes are wooded; the source of the Little River is situated at the foot of its east-southeastern flank and Bradley Pond is located at the bottom of its west-northwestern one.

20. [Luccombe Hill (Easter Hill)], 1,398.62 feet/426.30 metres, SS 907426, a roundish-shaped peak on a wide, northeast-projecting spur of Dunkery Hill, approaching 1¼ miles to the west-southwest. It rises fairly steeply from a tributary valley of Horner Water to the north, fairly steeply and then quite gently from the head of Hollow Combe to the west-northwest and fairly steadily from Hanny Combe to the east. OS Maps shows its summit to be at the OS 1:25000 map 426-metre spot height. However, it doesn't have a relative height of fifty feet and is linked to its parent peak by a very high col. A series of Bronze Age burial cairns, including Robin How and Joaney How, lie on its summit area and flanks and there is a fair amount of woodland on its lower northwestern, northern and eastern slopes.

21. [Red Stone Hill], 1,384.84 feet/422.10 metres, SS 811397, a slightly elongated, flattish-topped peak on an east-projecting spur of Little Ashcombe, well over 1¾ miles to the west-northwest. It rises steeply and then gently from the Exe Valley to the north and east and fairly steadily from the Pennycombe Valley to the south. OS Maps shows its summit to be at the OS 1:25000 map 424-metre spot height. However, it fails to be an independent hill by only 8.01 feet and is linked to its parent peak by a high col. A disused quarry and an adit are situated on its north-northwestern flank; Penn Allotment is located on its east-northeastern one and there are one or two areas of woodland on its lower eastern slopes. The B 3223 runs across its southern flank, broadly from east to west.

22. Lype Hill, 1,384.51 feet/422.00 metres, SS 950371, a large, elongated, flattish-topped peak, which rises increasingly gently from a col to the north, a tributary valley of the Washford River (via Colly Hill) to the east and one of the River Quarme to the west and gently from a headwater dale of the Pulham River to the south. OS Maps shows its highest point to be on the top of a Bronze Age bowl barrow, by the OS 1:25000 trig point, which is marked as standing at 423 metres. A second barrow is situated on its summit area; Kennisham Hill and its wireless station are located on its south-southeastern flank; Lype Common is sited on its east-southeastern one; Quarme Hill is found on its west-southwestern one; there is a fair amount of woodland on its lower slopes and the source of the Pulham River can be seen at the foot of its west-southwestern one. The B 3224 runs across its southern and western slopes, broadly from east-southeast to west-northwest.

23. [Kittuck Meads], 1,378.28 feet/420.10 metres, SS 826429, a small, flat-topped peak on a north-northwest-projecting spur of Great Hill (and, more immediately, Hurdle Down), more than a mile to the east. It rises fairly steadily and then very gently from Madacombe to the north, very gently from the same dale to the east and gently from a col to the west. OS Maps shows its summit to be at the OS 1:25000 map 422-metre spot height. However, the land drops by only 12.79 feet to a very high col linking it to its parent peak.

24. [Bill Hill], 1,376.64 feet/419.60 metres, SS 718410, a small, flattish-topped peak on a southwest-projecting spur of Butter Hill (and, more immediately, Broad Mead), more than 1¼ miles to the north-northwest. It is the most westerly 1,000-foot hill in Somerset. It rises fairly steadily and then gently from the Barle Valley to the east, very gently from the head of the Old Close Valley to the south and steadily and then gently from the head of one of the tributary dales of the latter to the west. OS Maps shows its summit to be towards the northern end of the OS 1:25000 map 420-metre small contour ring, but the land drops by only 10.50 feet to a very high col linking it to its parent peak. The county boundary with Devon runs across its southern and western flanks, broadly from south-southeast to north-northwest, and the B 3358 does the same, essentially from east-southeast to west-northwest.

25. [Horsen], 1,368.44 feet/417.10 metres, SS 779370, a small, flattish-topped peak on an east-projecting spur of Five Barrows Hill, more than three miles to the west. It rises steeply and then gently from the Barle Valley to the east, increasingly gently from Great Woolcombe to the north and gently from a col to the south-southeast. OS Maps shows its summit to be approximately in the middle of the OS 1:25000 map 410-metre contour bulge. However, the land drops by only 7.22 feet to a very high col linking it to its parent peak.

26. (Treborough Hill), 1,350.39 feet/411.60 metres, ST 005350, a flat-topped peak on a very small plateau, which rises steeply and then gently from the Washford Valley to the north, steeply and then very gently from one of its tributary dales to the east and variably from Guerney Bottom to the south and a col (via Withiel Hill and Langham Hill) to the west-northwest. OS Maps shows its highest point to be just west of the OS map 1:25000 412-metre spot height. A trig point is situated approximately 50 yards northwest of its summit. The peak is unnamed on the OS 1:25000 and 1:50000 maps, but it seems appropriate to name it after Treborough Common, which occupies most of its northern slopes. The Bronze Age Wiveliscombe (bowl) Barrow is situated just south of its summit area and another one, Huish Champflower Barrow, lies on its east-southeastern flank. Brendon Hill is situated on its south-southeastern flank; Beverton Pond (the source of the River Tone) is located at the foot of its east-southeastern one; the source of the Washford River is sited on its north-northeastern one; Withiel Hill and Langham Hill are found on its west-northwestern one and most of its west-northwestern and northern slopes are wooded. The B 3224 and the remains of the West Somerset Railway run across its upper southern flank, broadly from east-southeast to west-northwest.

27. Thornemead Hill, 1,349.74 feet/411.40 metres, SS 808374, a large, elongated, flat-topped peak, which rises steeply and then gently from the White Valley to the west, steeply and then very gently from the Pennycombe Valley to the east, fairly steadily and then gently from the Barle Valley to the south and variably from a col to the north-northwest. OS Maps shows its summit to be approximately 75 yards southwest of the OS 1:25000 map 412-metre spot height. The peak is unnamed on the OS 1:25000 and 1:50000 maps, but is called Thornemead Hill on several outdoors websites. Two prehistoric cairns are situated on its south-southwestern flank; there are one or two small areas of woodland on its lower slopes and the medieval Landacre Bridge is sited at the foot of its south-southeastern flank.

28. [Prescott Down (Stone Down)], 1,339.90 feet/408.40 metres, SS 865396, an elongated, flat-topped peak on a southeast to east-curving spur of Honeycombe Hill, just over 1¼ miles to the north-northeast. It rises steeply and then gently from the Quarme Valley to the north, increasingly gently from the same dale to the east and variably from the Exe Valley to the south. OS Maps shows its summit to be approximately 100 yards north of the OS 1:25000 map sheepfold, within the large 400-metre contour bulge. However, it doesn't have a relative height of fifty feet and is linked to its parent peak by a very high col. Kitnor Heath is situated on its eastern flank and a narrow strip of woodland runs along its lowest east-northeastern slopes. The B 3224 crosses its southern flank, broadly from west-southwest to east-northeast.

29. [Great Tom's Hill], 1,339.24 feet/408.20 metres, SS 815433, an elongated, but narrow plateau, which rises fairly steeply and then very gently from Stowford Bottom to the north, increasingly gently from Long Combe to the west and variably from the same dale to the south. OS Maps shows its summit to be at the eastern edge of the OS 1:25000 map 409-metre spot height. However, it doesn't have a relative height of fifty feet and is an outlier of Great Hill (and, more immediately, Hurdle Down and Kittuck Meads), approaching 1¾ miles to the east, to which it's linked by a high col. Black Hill is situated at the northwestern end of its plateau; a prehistoric round barrow and a double stone row are located on its western flank and Manor Allotment is sited on its west-northwestern one.

30. [Black Hill], 1,335.63 feet/407.10 metres, SS 806440, an elongated plateau, which rises pretty steeply and then gently from the Badgworthy Valley to the west, increasingly gently from Land Combe to the north and a tributary dale of Badgworthy Water to the south and quite gently and then gently from Stowford Bottom to the east. OS Maps shows its summit to be approximately 50 yards southwest of the OS 1:25000 map 408-metre spot height. However, it doesn't have a relative height of fifty feet and is an outlier of Great Hill (and, more immediately, Hurdle Down, Kittuck Meads and Great Tom's Hill), well over 2¼ miles to the east-southeast, to which it's linked by a very high col. Most of its upper slopes are occupied by South Common; Manor Allotment is situated on its southwestern flank; Stowey Allotment, Stowey Ridge and Oldhay Ridge are located on its north-northeastern one and there are areas of woodland on some of its lower slopes.

31. [Brendon Hill], 1,325.79 feet/404.10 metres, ST 015335, an elongated, flat-topped peak, which rises quite gently and then gently from Middleton Bottom to the south and gently from Beverton Pond (the source of the River Tone) to the north-northeast and the infant Tone Valley to the east. OS Maps shows its highest point to be just west of the B 3190 (which crosses the middle of its summit area), at or near the OS 1:25000 map unmarked 404-metre spot height. However, it doesn't have a relative height of fifty feet and is an outlier of Treborough Hill, to the north-northwest, to which it's linked by a high col. A radio station is situated at its summit; two transmission masts are located on its summit area; the source of the River Haddeo is sited on its west-southwestern flank; there are a number of areas of woodland on its lower slopes and Clatworthy Reservoir lies at the foot of its southeastern ones.

32. [Withiel Hill], 1,319.88 feet/402.30 metres, SS 993351, an elongated, flattish-topped peak on a west-northwest-projecting spur of Treborough Hill. It rises quite steeply and then very gently from a tributary valley of the Washford River to the north, gently from one of the River Haddeo to the south and variably from a col (via Langham Hill) to the west. OS Maps shows its summit to be at the OS 1:25000 map 404-metre spot height. However, it doesn't have a relative height of fifty feet and is linked to its parent peak by a very high col. The prehistoric Leather (bowl) Barrow is situated at the top of its west-northwestern flank and there is a considerable amount of woodland on its northern slopes. The remains of the West Somerset Railway run across its southern and western flanks, broadly from east-southeast to west-northwest, and the B 3224 crosses its northern one, essentially in the same direction.

33. Withypool Hill, 1,307.41 feet/398.50 metres, SS 839344, a somewhat elongated, roundish-topped peak, which rises fairly steadily and then quite gently from the West Valley to the south, increasingly gently from the Barle Valley to the east and variably from the same dale to the north and one of its tributary valleys to the west. OS Maps shows its highest point to be on the top of a Bronze Age round barrow, at or very near to the OS 1:25000 map unclear 398-metre spot height. The remains of Withypool Stone Circle are situated on its upper southwestern flank; there are a number of disused quarries on its east-southeastern one and most of its lower eastern slopes are covered by Hayes Wood and King's Wood.

34. [Tarr Ball Hill], 1,299.54 feet/396.00 metres, SS 865428, a small, flattish-topped peak on a long, north-northeast-projecting spur of Honeycombe Hill, well over a mile to the south-southwest. It rises steeply and then pretty gently from Lang Combe to the east and quite steeply and then pretty gently from the Nutscale Valley to the west and north. OS Maps shows its summit to be at the OS 1:25000 map 396-metre spot height. However, it doesn't have a relative height of fifty feet and is linked to its parent peak by a very high col. A prehistoric cairn lies on its upper western flank; Nutscale Reservoir is situated at the foot of its northwestern one and the remains of a Bronze Age barrow are located on its north-northeastern one. Its lowest eastern and northeastern slopes are covered by Wilmersham Wood.

35. [Humber's Ball], 1,292.32 feet/393.90 metres, SS 831328, a wedge-shaped, flattish-topped peak on a south-projecting spur of Brightworthy Hill, well over 1½ miles to the north-northwest. It rises steeply and then very gently from the Barle Valley (via Parsonage Down and Old Barrow Down) to the east-southeast, increasingly gently from a tributary dale of Dane's Brook to the west and variably from the West Valley to the east and Dane's Valley (via Fox Combe) to the south. OS Maps shows its summit to be at the OS 1:25000 map 394-metre spot height. However, it doesn't have a relative height of fifty feet and is linked to its parent peak by a very high col. A small wood is situated at the southern end of its summit area, Hawkridge Plain on its southwestern flank, Westwater Allotment on its east-northeastern one and Old Barrow Plantation on its eastern one.

36. (Elworthy Hill), 1,289.04 feet/392.90 metres, ST 070338, a very small plateau, which rises increasingly gently from a headstream of Halse Water to the east, gently from Holcombe to the west and variably from the Pond Valley (via Round Hill) to the north and Combe Davey to the south. OS Maps shows its highest point to be just southeast of the OS 1:25000 map 394-metre spot height. An unfinished Iron Age hillfort (Elworthy Barrows), a wireless station and a trig point (marked as standing at 392 metres) are situated on its summit area and there is a considerable amount of woodland on its flanks. The peak is unnamed on the OS 1:25000 and 1:50000 maps, but it seems reasonable to name it after the hillfort and the nearby Elworthy Combe. The B 3224 runs across the northern end of its summit area, broadly from east to west.

37. Road Hill, 1,281.50 feet/390.60 metres, SS 852370, a somewhat elongated, flat-topped peak, which rises steeply and then very gently from the Exe Valley to the east, increasingly gently from one of its tributary dales to the north, steeply and then gently from the Pennycombe Valley to the west, fairly steadily from the Barle Valley to the south and very gently from a col to the west-northwest. OS Maps shows its highest point to be at the OS 1:25000 map trig point, which is marked as standing at 391 metres. The Bronze Age Hernes (bowl) Barrow is situated approximately 50 yards to the south of its summit and the remains of the Iron Age Road Castle hillfort lie on its east-northeastern flank. Room Hill is located on its south-southeastern one; there is a fair amount of woodland on its lower slopes and the B 3223 runs across its southern and western ones, broadly from south-

southeast to north-northwest.

38. [Old Barrow Down], 1,274.28 feet/388.40 metres, SS 840324, a small, flattish-topped peak on an east-southeast-projecting spur of Brightworthy Hill (and, more immediately, Humber's Ball), approaching 2¼ miles to the north-northwest. It rises fairly steeply and then gently from Dane's Valley to the south and variably from the Barle Valley (via Parsonage Down) to the east and a tributary valley of West Water (via Westwater Allotment) to the north. OS Maps shows its summit to be at the OS 1:25000 map 388-metre spot height, on the top of the Bronze Age Old (bowl) Barrow. However, it doesn't have a relative height of fifty feet and is linked to its parent peak by a very high col. Clogg's Down is situated on its south-southwestern flank; Old Barrow Plantation is located on its northwestern one; Tarr Steps Woodland National Nature Reserve is sited on its lower eastern one and the famous Tarr Steps clapper bridge crosses the River Barle at its foot. Its lower eastern and east-southeastern slopes are heavily wooded.

39. Lyncombe Hill, 1,264.44 feet/385.40 metres, SS 875371, an elongated, flattish-topped peak, which rises steeply and then gently from the Exe Valley to the west and south, steeply and then very gently from the Larcombe Valley to the east and variably from the latter to the north. OS Maps shows its summit to be just east-northeast of the OS 1:25000 map 386-metre spot height. Staddon Hill is situated on its eastern flank; Staddon Hill Camp, an Iron Age hill slope enclosure, is located on its east-northeastern one and there are areas of woodland on its lower slopes.

40=. [Draydon Knap], 1,262.14 feet/384.70 metres, SS 893328, a somewhat elongated, flattish-topped peak, which rises fairly steeply and then gently from the Little Valley (via Contest Plantation) to the west, fairly gently from a tributary dale of the River Exe to the east and variably from Yellowcombe to the north. OS Maps shows its summit to be at the OS 1:25000 map 384-metre spot height. However, it fails to be an independent peak by just 0.13 feet and instead is an outlier of Winsford Hill, well over 1¼ miles to the northwest, to which it's linked by a high col. The Allotment and one or two disused quarries are situated on its greater summit area and several prehistoric remains are sited on its flanks. Its lower northern and northeastern slopes are wooded; the Caratacus Stone, an early medieval memorial stone, is located near the foot of its north-northwestern flank and the source of the Little River is found at the bottom of its northwestern one. The B 3223 runs across its upper western slopes, broadly from south to north.

40=. Wills Neck (Bagborough Hill), 1,262.14 feet/384.70 metres, ST 164351, an elongated peak, which rises variably from the Doniford Valley to the west, a col to the north, the Aisholt Valley (via Aisholt Common) to the east and West Bagborough to the south-southeast. OS Maps shows its highest point to be at the OS 1:25000 map 386-metre spot height, on the top of the middle of three Bronze Age round cairns situated on its summit area just northwest of the trig point, which is marked on the map as standing at 384 metres. Middle Hill is located on its east-southeastern flank; Black Knap and The Slades are sited on its east-northeastern one; the large, disused Triscombe Quarry is found on its northwestern one and there is a considerable amount of woodland on its slopes, particularly to the northeast. The A 358 runs across its western flank, broadly from south-southeast to north-northwest.

42. [Room Hill], 1,257.87 feet/383.40 metres, SS 860360, a somewhat elongated, flattish-topped peak on an east-southeast-projecting spur of Road Hill, to the north-northwest. It rises steeply and then gently from the Exe Valley to the east and one of its tributary dales to the north, increasingly gently from one of the River Barle to the west and gently from a col to the south-southeast. OS Maps shows its summit to be approximately in the middle of the OS 1:25000 map 380-metre contour ring. However, it doesn't have a relative height of fifty feet and is linked to its parent peak by a very high col. Curr Cleeve is situated on its northeastern flank; its lower eastern and northeastern slopes are fairly wooded and the B 3223 runs across its southern and western ones, broadly from south-southeast to north-northwest.

43. Winstitchen, 1,252.95 feet/381.90 metres, SS 787384, an elongated, flat-topped peak, which rises steeply and then gently from the Barle Valley to the west and south and the White Valley to the east and very gently from a col to the north-northwest. OS Maps shows its summit to be approximately 35 yards east of the OS 1:25000 map 382-metre spot height. The remains of Wheal Eliza, a disused copper and iron mine, are situated on its lower southwestern flank.

44. Monkham Hill, 1,250.00 feet/381.00 metres, SS 986393, an elongated, flat-topped peak, which rises increasingly gently from the Washford Valley to the south and the Pill Valley to the east, gently from a col to the west and variably from Long Combe (via Black Hill) to the north. OS Maps shows its summit to be approximately in the middle of the OS 1:25000 map 380-metre contour ring, about 125 yards south of the trig point, which is marked as standing at 381 metres. Various prehistoric remains are situated on its slopes, including those of two bowl barrows and earthworks of an Iron Age enclosure. Withycombe Common covers most of its upper slopes; Rodhuish Common is situated on its east-northeastern flank; the source of the Pill River is located on its east-southeastern one and there is a considerable amount of woodland on its lower slopes.

45. Croydon Hill, 1,248.03 feet/380.40 metres, SS 978392, a large, elongated, roundish-topped peak, which rises fairly steadily and then quite gently from a tributary dale of the River Avill to the north, fairly steadily and then gently from one of the Washford River to the south, very gently from a col to the east-northeast and variably from the Avill Valley to the north. OS Maps shows its highest point to be just west of the middle of the OS 1:25000 map 380-metre contour ring. The majority of the hill, including its summit area, is covered by woodland. Bat's Castle, an Iron Age hillfort, is situated on a subsidiary summit approaching ¾ mile to the north-northwest and Long Wood Enclosure, another Iron Age fort, is located on the peak's eastern flank. The village of Churchtown is situated on its south-southwestern flank; Timberscombe Common is located on its north-northwestern one; the village of Timberscombe is sited just below it and Longcombe Hill is found on its east-northeastern one.

46. [Riscombe Down], 1,241.47 feet/378.40 metres, SS 825393, a small peak on an east-projecting spur of Little Ashcombe (and, more immediately, Red Stone Hill), just over 2¾ miles to the west-northwest. It rises steeply and then gently from the Exe Valley to the north, increasingly gently from one of its tributary dales to the east and variably from the Pennycombe Valley to the south. OS Maps shows its summit to be approximately 60 yards northeast of the OS 1:25000 379-metre spot height. However, it doesn't have a relative height of fifty feet and is linked to its parent peak by a very high col. The B 3223 runs across the southern side of its summit area, broadly from east-southeast to west-northwest.

47. Little Cornham, 1,232.94 feet/375.80 metres, SS 757390, an ovalish-shaped peak, which rises steeply and then very gently from Cornham Brake to the east and south, fairly steadily and then quite gently from the Bale [sic] Valley to the north and quite gently from a col to the west. OS Maps shows its summit to be approximately on the letter "C", in the word "Cornham", as printed on the OS 1:25000 map towards the northern end of the 370-metre contour ring. There are several small areas of woodland on its lower slopes.

48. Mounsey Hill/Varle Hill, 1,227.03 feet/374.00 metres, SS 886316, an elongated, flattish-topped peak, which rises variably from the Barle Valley (via Great Common) to the south, steeply from the same dale to the west, fairly steadily and then quite gently from the Little Valley to the north, gently from the head of a tributary dale of the River Barle to the east and very gently from a col to the northeast. OS Maps shows its summit to be just east of the OS 1:25000 map 374-metre spot height. The peak is known by two names, with Mounsey Hill normally being used for its southern part and Varle Hill for its northern. The remains of Mounsey Castle, an Iron Age hillfort, are situated on a spur at the end of its southern flank; Ashway Side is located on its east-northeastern one and many of its lower slopes are wooded.

49. Chibbet Hill, 1,219.49 feet/371.70 metres, SS 838380, a short, flat-topped ridge, which rises quite steeply and then steadily from a tributary dale of Pennycombe Water to the south, increasingly gently from the Pennycombe Valley to the west and gently from a col to the north-northwest and the Exe Valley to the east. OS Maps shows its highest point to be at the OS 1:25000 map 372-metre spot height. The B 3223 runs across its summit area, broadly from southeast to northwest.

50. [North Hill], 1,212.27 feet/369.50 metres, SS 957342, an elongated, roundish-topped peak on a south-projecting spur of Lype Hill (and, more immediately, Kennisham Hill), well over 1¾ miles to the north-northwest. It rises fairly steadily and then gently from a tributary dale of the Pulham River to the west, quite gently and then gently from another one to the east and gently and then very gently from a col to the south. OS Maps shows its summit to be approximately at the OS 1:25000 map 371-metre spot height. However, it doesn't have a relative height of fifty feet and is linked to its parent peak by a high col. Although it is unnamed on the OS 1:25000 and 1:50000 maps, the peak immediately to the south of it is South Hill and therefore, by definition, this top must be North Hill. There is a fair amount of woodland on its lower southeastern slopes.

51. [Staddon Hill], 1,210.30 feet/368.90 metres, SS 886370, a small, flattish-topped peak on an east-southeast-projecting spur of Lyncombe Hill, to the west. It rises steeply and then gently from the Larcombe Valley to the north and east and the Exe Valley to the south. OS Maps shows its summit to be at the OS 1:25000 map 369-metre spot height. However, it doesn't have a relative height of fifty feet and is linked to its parent peak by a very high col. Staddon Hill Camp, an Iron Age hill slope enclosure, is situated on its north-northwestern flank and there are areas of woodland on the peak's lower slopes.

52. Bye Hill, 1,200.79 feet/366.00 metres, SS 881357, a flattish-topped peak on a short, east to west-running ridge. It rises steeply from the Exe Valley to the north and east, quite steeply and then fairly steadily from the Winn Valley to the south and gently from a col to the west-southwest. OS Maps shows its highest point to be at the western edge of the OS 1:25000 366-metre spot height. Its eastern outlier, approaching a mile away, is named on the OS 1:25000 map as Bye Hill, but its actual summit is on its western top, as noted above. There is a fair amount of woodland on its lower west-northwestern and eastern slopes and the village of Winsford is situated at the foot of its east-southeastern flank.

53. Lydeard Hill, 1,189.30 feet/362.50 metres, ST 179341, a wedge-shaped peak, which rises quite steeply and then fairly steadily from Grub Bottom to the south, quite steeply and then gently from the head of a tributary dale of the Black Stream to the west and variably from the Aisholt Valley to the north and the Peart Valley to the east. OS Maps shows its summit to be just east of the OS 1:25000 map 364-metre spot height. Two Bronze Age bowl barrows and a round cairn are situated on its greater summit area and there is a fair amount of woodland on its slopes.

54. [South Hill], 1,187.99 feet/362.10 metres, SS 899317, a small, flattish-topped peak on an east-projecting spur of Mounsey Hill/Varle Hill, to the west. It rises quite steeply and then very gently from a tributary dale of the River Exe to the east, fairly steadily and then gently from another to the north and variably from one of the River Barle to the south. OS Maps shows its summit to be at the OS 1:25000 map 361-metre spot height. However, it doesn't have a relative height of fifty feet and is linked to its parent peak by a very high col. Its lower eastern slopes are heavily wooded and the B 3223 runs across its southern and western ones, broadly from southeast to northwest.

55. Kersham Hill, 1,186.68 feet/361.70 metres, SS 948380, a somewhat elongated, flat-topped peak, which rises fairly steadily from Hart Cleeve to the south, increasingly gently from the Putham Valley to the west and a tributary dale of the Washford River to the east, gently from a col to the south-southeast and variably from a tributary valley of the River Avill to the north. OS Maps shows its summit to be towards the northeastern end of the OS 1:25000 map 360-metre contour ring, just north of the junctions of field boundaries. There are one or two disused quarries on its slopes and there is a fair amount of woodland on its lower ones.

56. Blagdon Hill, 1,182.09 feet/360.30 metres, SS 971337, a large, elongated, roundish-topped peak, which rises fairly steeply and then very gently from a tributary valley of the River Haddeo to the east and one of the Pulham Valley to the west, steadily and then gently from a col to the north-northeast and variably from Wimbleball Lake to the south. OS Maps shows its summit to be just east-northeast of the OS 1:25000 map 361-metre spot height. A trig point is situated on its summit area and is marked as standing at 360 metres; part of Kings Brompton Forest is located on its lower south-southwestern flank and Hurscombe Plantation is sited at the foot of its southern one.

57. [Langham Hill], 1,176.18 feet/358.50 metres, SS 980354, a small, flattish-topped peak on a west-northwest-projecting spur of Treborough Hill (and, more immediately, Withiel Hill), more than 1½ miles to the east-southeast. It rises very gently from a col to the west, quite gently and then gently from a tributary valley of the River Haddeo to the south and variably from one of the Washford River to the north. OS Maps shows its summit to be at the OS 1:25000 map 359-metre spot height. However, it doesn't have a relative height of fifty feet and is linked to its parent peak by a very high col. The B 3224 and the remains of the West Somerset Railway run across the middle of its summit area, broadly from east to west, and those of Langham Hill Pit are situated at the top of its west-northwestern flank. There is a large amount of woodland on its northern slopes.

58. (Hawkridge Hill) 1,170.93 feet/356.90 metres, SS 852312, a wedge-shaped, flattish-topped peak, which rises steeply and then gently from the Barle Valley to the east, steeply and then very gently from Dane's Valley to the south, increasingly gently from a tributary dale of the former to the north, quite gently and then very gently from one of the latter to the west and very gently from a col to the west-northwest. OS Maps shows its summit to be just south of the OS 1:25000 358-metre spot height. The peak is unnamed on the OS 1:25000 and 1:50000 maps, but it seems appropriate to name it after Hawkridge Common, Hawkridge Cross and the village of Hawkridge, all of which are situated on its slopes and may have been named after the hill! Hawkridge Ridge [sic] is located on its east-southeastern flank; disused quarries are sited on its north-northwestern one and there is a considerable amount of woodland on its lowest northeastern, eastern and east-southeastern slopes.

59. Black Hill, 1,168.96 feet/356.30 metres, ST 148380, a somewhat elongated, flat-topped peak, which rises fairly steadily from a col to the north and variably from another to the south-southeast, a dry valley (via Dead Woman's Ditch, a Bronze Age cross-dyke) to the east and Halsway Combe (via Hurley

Beacon) to the west. OS Maps shows its highest point to be at the OS 1:25000 map trig point, which is marked as standing at 358 metres. The remains of a Bronze Age cemetery, consisting of four bowl barrows and a cairn, lie on its summit area and there is a considerable amount of woodland on its lower slopes. Wilmot's Pool is situated on its eastern flank; Frog Hill and Robin Upright's Hill are located on its east-northeastern one and there are a number of disused quarries on its slopes. The A 358 runs across its west-southwestern flank, broadly from south-southeast to north-northwest.

60. Ison Hill, 1,165.35 feet/355.20 metres, SS 904369, a roundish, flattish-topped peak, which rises steeply and then very gently from the Quarme Valley to the east and the Larcombe Valley to the west and fairly steadily and then gently from their tributary dales to the north and south, respectively. OS Maps shows its summit to be approximately 50 yards northwest of the OS 1:25000 map 355-metre spot height. Although the peak is unnamed on the OS 1:25000 and 1:50000 maps, its traditional name is Ison (or Eisen) Hill and it is sometimes called Oldrey Hill. There are a number of small areas of woodland on its lower slopes.

61. Haddon Hill (Hadborough), 1,158.14 feet/353.00 metres, SS 961286, an oval-shaped peak, the highest on a 3½-mile-long, east to west-running ridge. It rises steeply and then fairly steadily from Haddeo Valley to the north, variably from the same dale to the west and one of its tributary valleys to the south and very gently and then gently from a col to the east. OS Maps shows its summit (named Hadborough) to be just northeast of the OS 1:25000 map trig point, which is marked as standing at 355 metres. Its lower western, northwestern and northern slopes are heavily wooded; Hadborough Plantation stretches across its upper southern flank and Wimbleball Lake is situated at the foot of its northeastern one. The B 3190 runs across its southeastern and eastern slopes, broadly from south-southwest to north-northeast.

62. [Black Knap], 1,155.18 feet/352.10 metres, ST 171354, a small peak on a narrow, east-projecting spur of nearby Wills Neck, to the west-southwest. It rises steeply from Cockercombe to the north and variably from the Aisholt Valley to the east and a dry dale to the south. OS Maps shows its summit to be near to the eastern end of the OS 1:25000 map 350-metre contour bulge, but the land drops by only 9.51 feet to a very high col linking it to its parent peak. Its summit area and northern, northeastern and upper eastern slopes are heavily wooded; Aisholt Common is situated on its east-southeastern flank; Round Hill is located on its east-northeastern one and The Slades are found on its north-northeastern one.

63. [Hurley Beacon], 1,154.20 feet/351.80 metres, ST 142380, a small, flattish-topped peak on the western end of the greater summit area of Black Hill. It rises steeply and then fairly steadily from a tributary valley of the Doniford Stream to the south and Halsway Combe (via a disused quarry) to the west and quite gently from a col the north-northeast. OS Maps shows its summit to be at the OS 1:25000 map 353-metre spot height. However, it doesn't have a relative height of fifty feet and is linked to its parent peak by a very high col. A Bronze Age round cairn, which has the same name as the peak itself, is one of three lying on its summit area and Crowcombe Park is situated on its southern flank. There is a fair amount of woodland on its lower southern slopes; the village of Crowcombe is located on its south-southeastern flank and the A 358 runs across its west-southwestern one, broadly from south-southeast to north-northwest.

64. Exton Hill, 1,144.36 feet/348.80 metres, SS 935346, an elongated, roundish-topped peak, which rises fairly steeply and then very gently from a tributary dale of the River Exe to the south, increasingly gently from the Quarme Valley to the west, quite gently and then gently from the Pulham Valley to the east and gently from a col to the north. OS Maps shows its summit to be at the OS 1:25000 map 349-metre spot height. The village of Exton is situated on its southwestern flank and there are a number of wooded areas on its slopes. The A 396 runs along its lowest western ones, broadly from south to north.

65. (Ashcombe Hill), 1,140.09 feet/347.50 metres, SS 890379, an elongated peak, which rises steeply from the Larcombe Valley to the west and south and less so from the Quarme Valley to the east. OS Maps shows its summit to be just east of the OS 1:25000 map 349-metre spot height. The peak is unnamed on the OS 1:25000 and 1:50000 maps, but it seems appropriate to name it after Ashcombe, the nearest settlement to it, at the foot of its southwestern flank.

66. [Fire Beacon (West Hill)], 1,125.33 feet/343.00 metres, ST 153371, a somewhat elongated peak, which rises steeply and then gently from Crowcombe Combe to the west, quite steeply and then fairly gently from Little Quantock Combe to the south, fairly steadily and then quite gently from Rams Combe to the north, fairly steadily from a col to the southeast and variably from Quantock Combe (via Lord's Ball) to the east. OS Maps shows its summit to be at the OS 1:25000 map 342-metre spot height. However, it fails to be an independent peak by just 0.13 feet and instead is an outlier of Black Hill, to the north-northwest, to which it's linked by a high col. A Bronze Age round cairn is situated at the eastern end of its summit area; there are one or two disused quarries on its slopes and its lower eastern and northeastern ones are heavily wooded.

67. (Draydon Hill), 1,121.06 feet/341.70 metres, SS 894301, an elongated, flattish-topped peak, which rises steeply and then gently from the Barle Valley to the south and fairly steadily and then gently from its tributary dales to the east and west. OS Maps shows its summit to be at the OS 1:25000 map 341-metre spot height. The peak is unnamed on the OS 1:25000 and 1:5000 maps, but it seems appropriate to name it after Draydon Farm, the nearest feature, which is situated on its upper southern flank. Great Common is located on its west-southwestern flank and most of its lower slopes, except for its northern ones, are wooded.

68=. [Edbrooke Hill], 1,119.75 feet/341.30 metres, SS 909331, an elongated, twin-summited peak, which rises quite steeply and then gently from Yellowcombe to the north, variably from the Exe Valley to the east and quite steeply and then very gently from one of the tributary dales of the latter to the south. OS Maps shows its highest point to be on its southern top, approximately 90 yards south of the OS 1:25000 map 342-metre spot height. Its northern summit is approximately 400 yards away on the western side of the map's 340-metre small contour ring and is computed by OS Maps to be just 3.94 feet lower. However, Edbrooke Hill doesn't have a relative height of fifty feet and is an outlier of Winsford Hill (and, more immediately, Draydon Knap), well over two miles to the west-northwest. Most of its lowest slopes are heavily wooded.

68=. Great Hill, 1,119.75 feet/341.30 metres, ST 155365, a twin-summited peak, which rises steeply and then fairly steadily from Little Quantock Combe to the west, fairly steadily and then steeply from Triscombe Combe to the south, gently from a col to the north and variably from Cockercombe (via Floorey Down) to the east. OS Maps shows its highest point to be on its north-northwestern summit, at the OS 1:25000 map 339-metre spot height. Its south-southeastern (subsidiary) top is computed by OS Maps to be approximately 425 yards away, roughly in the middle of the OS 1:25000 map 335-metre very small contour ring, with a height of 1,101.38 feet. However, the OS map marks its summit as 337 metres and positions it on the top of a Bronze Age platform cairn. Marrow Hill is situated on its east-southeastern flank; its lower eastern and east-southeastern slopes are heavily wooded and the A 358 runs across its western flank, broadly from south-southeast to north-northwest.

70. Heydon Hill, 1,119.42 feet/341.20 metres, ST 035281, an elongated, flattish-topped peak, which rises steeply and then gently from the Tone Valley to the east, variably from one of its tributary dales to the north and a col to the south and gently and then quite gently from a another col to the west-northwest. OS Maps shows its highest point to be at the OS 1:25000 map 342-metre spot height. Two prehistoric bowl barrows are situated to the south-southeast of its summit; the village of Chipstable is located on its south-southeastern flank; Huish Moor lies on its north-northeastern one and there is a sizable amount of woodland on its slopes.

71. Couple Cross Hill, 1,119.09 feet/341.10 metres, SS 955387, an elongated, slightly rocky peak, which rises fairly steadily from a tributary dale of the Washford River to the south, fairly steeply and then quite gently from one of the River Avill to the east and variably from the head of another of the latter to the north and the Putham Valley to the west. OS Maps shows its summit to be approximately 35 yards east of the OS 1:25000 map 344-metre spot height, by a small quarry. The hill is unnamed on the OS 1:25000 and 1:50000 maps, but it has recently become named after the crossroads to the west-southwest, even though Stowey Hill seems a more appropriate title because Old Stowey, Stowey Farm and Stowey Plantation are all situated nearby. There are quite a number of small areas of woodland and several disused quarries on its slopes.

72. (Blagdon Hill), 1,117.45 feet/340.60 metres, SS 909395, an oval-shaped peak, which rises steeply and then fairly steadily from a tributary dale of the River Avill to the north, steadily and then gently from another to the east and fairly steadily from the Quarme Valley to the south. OS Maps shows its summit to be at the OS 1:25000 map 342-metre spot height. The hill is unnamed on the OS 1:25000 and 1:50000 maps, but it seems appropriate to name it after Blagdon Wood, which is situated on its lower northern and northeastern slopes; Blagdon Farm at the foot of its eastern flank and Blagdon Cross on its east-southeastern one. The B 3224 runs across its southern slopes, broadly from west-southwest to east-northeast.

73. Thorncombe Hill, 1,116.47 feet/340.30 metres, ST 131391, the highest peak on an elongated, snake-shaped, small plateau. It rises steeply and then fairly steadily from Slaughterhouse Combe to the east, steeply and then quite gently from Paradise Combe to the west, quite gently and then very gently from a col to the north-northwest, and variably from Halsway Combe to the south. OS Maps shows its highest point to be just north-northwest of the OS 1:25000 map 341-metre spot height. Halsway Hill is situated on its south-southeastern flank; Black Ball Hill is located on its north-northeastern one; Bicknoller Hill and the large Bronze Age Thorncombe (bowl) Barrow are sited on its west-northwestern one; a number of other prehistoric remains are found on its slopes and its lower eastern and northeastern ones are wooded. The A 358 runs across its southwestern flank, broadly from east-southeast to west-northwest.

74. Combeshead Hill (Picket Hill), 1,108.92 feet/338.00 metres, SS 943327, a small plateau, which rises steeply and then gently from the Pulham Valley to the east, fairly steeply and then quite gently from a tributary dale of the River Exe to the west and gently and then very gently from one of the former to the south. OS Maps shows its summit to be approximately on the letter "r", in the word "Cross", as printed on the OS 1:25000 map, towards the northern end of the 330-metre large contour ring. The village of Brompton Regis is situated on its south-southeastern flank and its lowest west-southwestern and east-northeastern slopes are wooded. The A 396 snakes along the foot of its southwestern flank, broadly from south to north.

75. Dowsborough, 1,107.61 feet/337.60 metres, ST 161391, an elongated, ovalish-topped peak, which rises steeply and then fairly steadily from Bin Combe to the east, steeply and then gently from Holford Combe to the west and variably from a col to the south and an unnamed valley to the north. OS Maps shows its highest point to be at the OS 1:25000 map 339-metre spot height. The remains of Dowsborough Camp, a sizable Iron Age hillfort, and a prehistoric round barrow are situated on its summit area. Almost all of its slopes, except those to the north, and its summit area are covered by woodland. The site of Walford's Gibbet is situated on its east-northeastern flank; Knacker's Hole spring and Woodlands Hill are located on its north-northwestern one and there are one or two disused quarries on its slopes. The A 39 runs across its east-northeastern flank, broadly from east-southeast to west-northwest.

76. [Great Ferny Ball], 1,106.96 feet/337.40 metres, SS 797366, a small peak on a north-northeast-projecting spur of Horsen Hill, to the south-southwest. It rises quite steeply from the Barle Valley to the east and steeply from the same dale to the north and one of its tributary valleys to the west. OS Maps shows its summit to be approximately in the middle of the northern end of the OS 1:25000 map 330-metre contour bulge. However, the hill doesn't have a relative height of fifty feet and is linked to its parent peak by a very high col.

77. [Halsway Hill], 1,104.66 feet/336.70 metres, ST 132387, a small, flattish-topped peak on a south-southeast-projecting spur of nearby Thorncombe Hill. It rises steeply and then fairly steadily from Halsway Combe to the south and Paradise Combe to the west and variably from Stert Combe to the east. OS Maps shows its summit to be approximately in the middle of the OS 1:25000 map 335-metre small contour ring, but the land drops by only 7.22 feet to a very high col linking it to its parent peak. Its lower northeastern slopes are heavily wooded and the A 358 runs across its west-northwestern flank, broadly from east-southeast to west-northwest.

78. Flexbarrow, 1,103.02 feet/336.20 metres, SS 782381, a very small, slightly elongated peak, which rises steeply from the Barle Valley to the east, south and west and from a trough to the north. OS Maps shows its summit to be at the northern edge of the OS 1:25000 map 339-metre spot height.

79. (Little Haddon Hill), 1,102.69 feet/336.10 metres, SS 997283, an elongated, flat-topped peak, which rises steeply and then quite gently from the Batherm Valley to the east, variably from one of its tributary dales to the south and the Haddeo Valley to the north and very gently from a col (via Britannia's Shield plantation) to the west. OS Maps shows its summit to be approximately in the middle of the OS 1:25000 map 330-metre contour ring. The peak is unnamed on the OS 1:25000 and 1:50000 maps, but it seems appropriate to name it after Haddon End, on its south-southeastern flank, and Little Haddon Farm, on its south-southwestern one, especially because it's a part of the same upland as Haddon Hill, to the west. The village of Skilgate is situated on its southwestern flank; Upton Cleave is located on its west-northwestern one; Wimbleball Lake is sited at the foot of the latter and there is a fair amount of woodland on its slopes. The B 3190 runs across its northern flank, broadly from west-southwest to east-northeast.

80. [Beacon Hill], 1,091.21 feet/332.60 metres, ST 049358, a small peak on a north-northeast-projecting spur of Treborough Hill, more than 2¾ miles to the west-southwest. It rises fairly steeply and then quite gently from a tributary valley of the Washford River to the west and variably from one of the Doniford Stream to the east and a col to the north-northwest. OS Maps shows its summit to be at the OS 1:25000 map 334-metre spot height, by a small reservoir, but the land drops by only 12.47 feet to a very high col linking it to its parent peak. There are large areas of woodland on the slopes of its northern hemisphere; Bird's Hill and North Bird's Hill are situated on its east-northeastern flank and one or two disused quarries are located on its slopes. The B 3190 runs across its western flank, broadly from south-southwest to north-northeast.

81. Cothelstone Hill, 1,087.93 feet/331.60 metres, ST 189326, a somewhat elongated peak, which rises steeply and then gently from a tributary valley of the Back Stream to the west and variably from the same dale (via the hamlet of Toulton) to the south and cols to the east-northeast and north-northwest. OS Maps shows its rocky highest point to be at the OS 1:25000 map 332-metre spot height. A prehistoric bowl barrow and the remains of an eighteenth century tower are situated at its summit. Its summit area and upper flanks are almost surrounded by woodland; Dene Hill Brake is situated on its south-southeastern flank; Merridge Hill is located on its east-northeastern one; Cothelstone Park is sited on its west-southwestern one and the village of Cothelstone is found on its southwestern one.

82. (Furzehill), 1,087.60 feet/331.50 metres, SS 910356, a somewhat elongated, roundish-topped peak, which rises steeply and then gently from the Quarme Valley to the east, very gently from a col to the north and variably from the Exe Valley to the west and south. OS Maps shows its summit to be on the northern edge of the OS 1:25000 map 332-metre spot height. A trig point, marked on the map as standing at an altitude of 331 metres, is situated a short way to the west-southwest. The peak is unnamed on the OS 1:25000 and 1:50000 maps, but it seems appropriate to name it after Furzehill Lane, which climbs its lowest eastern slopes and may have taken the name of the peak in the first place! There is a fair amount of woodland on some of its lower slopes.

83. [Bicknoller Hill], 1,086.61 feet/331.20 metres, ST 127394, a small, somewhat elongated peak on a south-southwest and a west-projecting spur of nearby Thorncombe Hill, to the east-southeast. It rises steeply and then fairly steadily from Paradise Combe to the south and Long Combe to the west, fairly steeply and then quite gently from the head of Bicknoller Combe to the north and variably from a col to the west-northwest. OS Maps shows its summit to be at the OS 1:25000 map 332-metre spot height, on the top of the prehistoric Thorncombe (bowl) Barrow. However, it doesn't have a relative height of fifty feet and is linked to its parent peak by a very high col. Trendle Ring prehistoric hillfort settlement and a disused quarry are situated on its western flank and the village of Bicknoller is located at the foot of the same slopes.

84. (Treborough North Hill), 1,085.96 feet/331.00 metres, ST 008365, an elongated, flattish-topped peak, which rises variably from the Washford Valley to the north, increasingly gently from the same dale to the west and steadily from one of its headwater valleys to the south and east. OS Maps shows its summit to be towards the western side of the OS 1:25000 map 330-metre very small contour ring. The hill is unnamed on the OS 1:25000 and 1:50000 maps, but, as the Treborough place name appears aplenty in the vicinity and the peak is north of Treborough Hill, it seems appropriate to name it as above. There is a fair amount of woodland on its lower north-northeastern, northwestern and west-northwestern slopes and the village of Kingsbridge is situated at the foot of its west-northwestern flank.

85. [Marrow Hill (Hart Hill)], 1,078.08 feet/328.60 metres, ST 162361, a small, roundish-topped peak on a short, south-projecting spur of nearby Great Hill, to the west-northwest. It rises steeply and then fairly steadily from Two Tree Bottom and Cockercombe to the east, fairly steadily and then quite gently from Triscombe to the south and variably from Quantock Combe to the north. OS Maps shows its summit to be at the OS 1:25000 map 329-metre spot height. However, it doesn't have a relative height of fifty feet and is linked to its parent peak by a high col. The Triscombe Stone, a probable Bronze Age menhir, is situated on its southern flank and almost all of its northeastern, eastern and southeastern slopes are covered in woodland.

86. (Broadway Head), 1,074.15 feet/327.40 metres, ST 063321, a short, east-west-running ridge, which rises steeply and then very gently from Clatworthy Reservoir to the west-southwest, steeply and then fairly gently from the Hillfarrance Valley to the east, quite steeply and then gently from a tributary dale of the reservoir to the west, gently from a col to the north-northeast and variably from another to the south. OS Maps shows its summit to be at the OS 1:25000 map 328-metre spot height. The peak is unnamed on the OS 1:25000 and 1:50000 maps, but it seems appropriate to name it after the nearest settlement, Broadway Head Farm, on its northern flank. The village of Clatworthy is situated on its south-southwestern flank and there are areas of woodland on some of its lower slopes.

87. Storridge Hill, 1,072.83 feet/327.00 metres, SS 946302, a round peak, with a large, flattish top, which rises very steeply and then very gently from Hartford Bottom to the south, fairly steadily and then very gently from a dry valley to the west, quite gently and then very gently from a col to the west-northwest, increasingly gently from a tributary dale of the Pulham River to the east and variably from another to the north. OS Maps shows its summit to be approximately 80 yards west of the OS 1:25000 map 327-metre spot height. Its southern and lower southeastern and eastern slopes are heavily wooded.

88. [Little Heydon (Potter's Knap)], 1,068.24 feet/325.60 metres, ST 024286, an elongated, flat-topped peak on a west-northwest-projecting spur of Heydon Hill. It rises fairly steadily and then very gently from the Batherm Valley to the west, fairly steadily and then gently from one of its tributary dales to the south and variably from one of its headwater valleys to the north. OS Maps shows its summit to be at the OS 1:25000 map 326-metre spot height. However, it doesn't have a relative height of fifty feet and is linked to its parent peak by a high col. Oxenleaze Brake is situated on its south-southwestern flank; Heydon Common is located on its upper west-southwestern one; Lotley Brakes are sited on its western one and there is a fair amount of woodland on some of its slopes.

89. (Broford Hill), 1,067.59 feet/325.40 metres, SS 911306, an oval-shaped, flat-topped peak, which rises increasingly gently from a tributary dale of the River Barle to the west, very gently from cols to the south-southeast and north-northwest and variably from the Exe Valley to the east. OS Maps shows its summit to be at the OS 1:25000 map 326-metre spot height. The peak is unnamed on the

OS 1:25000 and 1:50000 maps, but it seems appropriate to name it after two nearby settlements and woods to its northeast and east-northeast, which contain the word "Broford". There are one or two disused quarries on its slopes and most of the lower ones of its eastern hemisphere are wooded.

90. (Cow Castle Hill), 1,065.62 feet/324.80 metres, SS 794373, a very small, isolated knoll, which rises steeply from a trough to the north, steeply and then fairly steadily from the White Valley to the east and steadily and then steeply from the Barle Valley to the south and west. OS Maps shows its summit to be approximately 50 yards east of the OS 1:25000 map 332-metre spot height and strangely *below* the map's 330-metre small contour ring! The sizable discrepancy in the peak's altitude given by the online and paper versions of the OS's data might be explained by a limitation of the website's programme to calculate accurately the heights of rapidly-rising features. The peak is unnamed on the OS 1:25000 and 1:50000 maps, but it's obviously appropriate to name it after Cow Castle, the Iron Age hillfort, which occupies its greater summit area.

91. Black Down (Beacon Batch), 1,059.06 feet/322.80 metres, ST 484573, a large, elongated peak, which is the most northerly 1,000-foot hill in Somerset. It rises steeply and then very gently from Rowberrow Bottom to the west, very gently from a col to the east-southeast and variably from Burrington Combe to the north and Rhino Rift to the south. OS Maps surprisingly shows its summit (named Beacon Batch) to be approximately 60 yards north of the OS 1:25000 map trig point, which tops a Bronze Age round barrow and is marked as standing at 325 metres. A Bronze Age barrow cemetery is situated on its summit area and the remains of a Second World War bombing decoy complex lie on the higher parts of the hill. Gorsey Bigbury Neolithic henge is located on its southern flank; Tyning's Farm Swallet is sited on its south-southwestern one; Rowberrow Warren and the Palaeolithic Rowberrow Cavern are found on its west-northwestern one; Read's Cavern, Bos Swallet, Rod's Pot, Sidcot Swallet, Goatchurch Cavern and Whitcombe's Hole are all positioned on its northwestern one; East Twin Swallet can be seen on its north-northwestern one and the source of the Cheddar Yeo River is perceived at the foot of its east-southeastern one. Its lower western and lowest southern and north-northeastern slopes are heavily wooded. The B 3134 runs along the foot of its northern flank, broadly from east-southeast to west-northwest.

92. [Broom Ball], 1,058.40 feet/322.60 metres, SS 869385, a small conical peak on a short, west-southwest-projecting spur of Lyncombe Hill, approaching a mile to the south-southeast. It rises fairly steadily from a tributary valley of the River Exe to the south, increasingly gently from another to the west and gently from yet another to the north. OS Maps shows its summit to be just east-northeast of the OS 1:25000 map 325-metre spot height. However, it doesn't have a relative height of fifty feet and is linked to its parent peak by a very high col.

93. [Higher Hare Knap], 1,053.15 feet/321.00 metres, ST 150391, the highest point on a 1¾-mile-long, curving ridge. It rises very steeply and then fairly steadily from Holford Combe to the east and Stert Combe to the west and variably from Hodder's Combe to the north. OS Maps shows its summit to be approximately in the middle of the OS 1:25000 map long, narrow 320-metre contour ring. The small rise, with the 312-metre spot height and a ruined prehistoric cairn, about 440 yards to the north-northwest, is usually regarded as the summit of the Knap, but its true top is that noted above. It fails to be an independent hill by just 5.38 feet and instead is an outlier of Black Hill, to the south-southwest, to which it's linked by a high col. There are several prehistoric remains on its slopes; Lower Hare Knap is situated on its north-northwestern flank and its lower eastern and western slopes are heavily wooded.

94. Ley Hill (Doverhay Down), 1,045.93 feet/318.80 metres, SS 886446, a somewhat elongated, roundish-topped peak, which rises steeply and then quite gently from the Horner Valley to the east and south and Hawk Combe to the north and steeply and then very gently from the Nutscale Valley to the west. OS Maps shows its summit to be just east of the OS 1:25000 map 318-metre spot height. The remains of two Bronze Age bowl barrows and a probable medieval enclosure are situated on its northern flank and Crawter Hill is located on its north-northeastern one. Most of its lower southern slopes are covered by Wilmersham Wood, its eastern ones by Horner Wood and its northern ones by Doverhay Plantation.

95=. [Robin Upright's Hill], 1,040.03 feet/317.00 metres, ST 163383, a small, ovalish-shaped peak on an east-northeast to north-northwest-curving spur of Black Hill, a mile to the west-southwest. It rises fairly steadily and then gently from a col to the north-northwest and variably from an unnamed valley to the east and Rams Combe to the south. OS Maps shows its summit to be at the OS 1:25000 map 318-metre spot height, but the land drops by only 6.57 feet to a very high col linking it to its parent peak. Great Bear and a number of disused quarries are situated on its east-northeastern flank; Lady's Fountain spring is located on its west-northwestern one and most of its slopes are wooded.

95=. (Woodcocks Hill), 1,040.03 feet/317.00 metres, SS 876450, a wedge-shaped, roundish-topped peak, which rises steeply and then fairly steadily from Hawk Combe to the north and the Nutscale Valley to the south, fairly steadily from Ley Combe to the east and quite gently from a col to the west. OS Maps shows its summit to be at the OS 1:25000 map 317-metre spot height. The hill is unnamed on the OS 1:25000 and 1:50000 maps, but it seems appropriate to name it after the nearest settlement to it, Woodcocks Ley, which is situated on its lower eastern flank. Hawkcombe Woods cover most of its lower northern slopes and Wilmersham Wood stretches along its lower southern ones.

97=. [Court Down], 1,036.42 feet/315.90 metres, SS 914298, a very short, east to northwest-curving ridge, which rises variably from the Barle Valley and Dulverton to the south, steeply and then fairly steadily from one of its tributary dales to the west and steeply and then gently from the Exe Valley to the east. OS Maps shows its summit to be approximately 190 yards northwest of the OS 1:25000 map trig point, which is marked as standing at 316 metres. However, the hill doesn't have a relative height of fifty feet and is an outlier of Broford Hill, to the north-

northwest, to which it's linked by a high col. Hele Ball is situated on its lower south-southeastern flank; most of its lower slopes are wooded and the B 3223 cuts across its south-southeastern and southern ones and runs along its lowest south-southwestern and western ones, broadly from southeast to northwest.

97=. Longstone Hill, 1,036.42 feet/315.90 metres, ST 136404, an elongated, flattish-topped peak, with a 2½-mile-long northern spur to the Bristol Channel. It rises very steeply and then quite gently from Sheppard's Combe (via Lady's Edge) to the south and Hodder's Combe to the east, increasingly gently from Weacombe Combe to the west, very gently from a col to the west-northwest and variably from Bridgwater Bay and Quantock's Head to the north. OS Maps shows its summit to be approximately in the middle of the OS 1:25000 map 315-metre contour ring. There are a number of prehistoric remains on its flanks, probably including the Long Stone to the east, which was also an ancient boundary stone. Pardlestone Hill is situated on its northeastern flank; Broom Ball is located on its north-northeastern one and its lowest eastern and southeastern slopes are heavily wooded. The A 39 runs across its northern flank, broadly from east to west.

97=. West Hill (South Hill), 1,036.42 feet/315.90 metres, SS 978299, a wedge-shaped, flattish-topped peak, which rises steeply and then fairly steadily from Wimbleball Lake to the south, fairly steeply and then steadily from the same reservoir to the north and west and variably from a tributary valley of the River Haddeo (via East Moor and Great Moor) to the east. OS Maps shows its summit to be approximately 80 yards northwest of the more northerly junction of field boundaries within the OS 1:25000 map 310-metre contour ring. The peak is known both as West Hill and South Hill, as per West Hill Wood and South Hill Wood on its lower western and southern slopes, respectively, but it seems that the former name is the better known.

100. [South Hill], 1,035.76 feet/315.70 metres, SS 958325, an elongated, twin-summited peak on a long, south-projecting spur of Lype Hill (and, more immediately, North Hill), approaching three miles to the north. It rises steeply and then gently from the Pulham Valley to the west and one of its tributary dales to the east and fairly steadily and then quite gently from the former to the south. OS Maps shows its highest point to be on its northern summit, approximately 45 yards south of the OS 1:25000 map 316-metre spot height. Although its southern top is marked as reaching 317 metres at its spot height, OS Maps computes its highest point to be about 55 yards to the northwest and 0.33 feet lower than the northern summit. The hill doesn't have a relative height of fifty feet and is linked to its parent peak by a high col. South Hill Wood and part of Kings Brompton Forest cover almost all its lower eastern, southeastern and south-southeastern slopes.

101. Staple Hill, 1,032.81 feet/314.80 metres, ST 240166, a large, elongated, ovalish-topped peak, which is the most southerly 1,000-foot hill in Somerset. It rises quite steeply and then very gently from the headwaters of the Sherford Stream to the west, fairly steadily from the Broughton Valley to the north and variably from a headwater stream of the River Yarty to the south and an infant dale of the Fivehead River to the east. OS Maps shows its summit to be just east of the OS 1:25000 map trig point, which is marked as standing at 315 metres. Its summit area and the majority of its slopes, except to the south, are heavily wooded and North Down is situated on its west-southwestern flank. The B 3170 runs across its western and northern slopes, broadly from south-southwest to north-northeast.

102. Lyddon's Hill, 1,030.84 feet/314.20 metres, SS 937295, a narrow, south-southwest to north-northeast-running, flattish-topped ridge, which rises steeply and then fairly steadily from the Exe Valley to the west, quite gently and then very gently from a col to the south and variably from the head of one of its tributary dales to the north and a dry valley to the east. OS Maps shows its summit to be just west-southwest of the OS 1:25000 map 314-metre spot height. Butter Ball is situated on its south-southeastern flank and its lower slopes, except in the north, are almost entirely covered by woodland.

103. (Charterhouse Hill), 1,022.64 feet/311.70 metres, ST 498568, an ovalish-shaped, flattish-topped peak, which is the most easterly 1,000-foot hill in Somerset. It rises gently and then fairly gently from a dry dale to the east, very gently from a col to the west-northwest and variably from the Yeo Valley (via Blagdon) to the north and Velvet Bottom to the south. OS Maps shows its summit to be at the eastern end of the OS 1:25000 map 315-metre very small contour ring, approximately 60 yards north of the map's 311-metre spot height. The hill is unnamed on the OS 1:25000 and 1:50000 maps, but it seems appropriate to name it after the nearby hamlet of Charterhouse, which is situated on its southern flank. A wireless station and a small reservoir are located on its summit area; the remains of a Roman lead-mining settlement and fort are sited on its south-southeastern flank; there are (mainly small) areas of woodland on its slopes and the source of the Cheddar Yeo River is found at the foot of its western flank. The B 3134 and the A 368 run across its northern slopes, broadly from east-southeast to west-northwest.

104. (Whitefield Hill), 1,017.39 feet/310.10 metres, ST 064303, an elongated peak, which rises steeply and then quite gently from a tributary dale of the Hillfarrance Brook to the east, quite steeply and then gently from the Tone Valley to the west and variably from the Blackwater Valley to the south and a col to the north. OS Maps shows its summit to be just south-southwest of the OS 1:25000 map 311-metre spot height and its trig point, a little to the east-southeast, to be only 0.33 feet lower, although the map marks it as standing at 309 metres. The peak is unnamed on the OS 1:25000 and 1:50000 maps, but it seems appropriate to name it after the village of Whitefield, on its east-southeastern flank, by which name it's already known to some extent. Woodcock Brake is situated on its northeastern flank and most of its lowest eastern slopes are wooded.

105. [Beacon Hill], 1,017.06 feet/310.00 metres, ST 124409, an ovalish-shaped, flattish-topped peak on a west-northwest to north-curving spur of Longstone Hill, to the east-southeast. It rises very steeply and then fairly steadily from Weacombe Combe to the west, steeply and then gently from the same dale to the south and variably from Bridgwater Bay (via West Hill) to the north and

Gay's House Combe to the east. OS Maps shows its summit to be approximately 60 yards south-southeast of the OS 1:25000 map trig point, which is marked as standing at 310 metres. However, it doesn't have a relative height of fifty feet and is linked to its parent peak by a high, narrow col. A number of prehistoric platform cairns are situated on its summit area; the village of West Quantoxhead is sited on its west-northwestern flank; Stowborrow Hill is located on its north-northwestern one; Round Hill is found on its north-northeastern one and there are several areas of woodland on its slopes. The A 39 runs across its northwestern and northern flanks, broadly from west-southwest to east-northeast.

106. (Northmoor Hill), 1,009.19 feet/307.60 metres, SS 891287, an elongated, wedge-shaped peak, which rises variably from the Barle Valley to the east, steeply and then gently from the same dale to the north, quite steeply and then gently from a headwater valley of the Brockey River to the south and gently from a col to the west-southwest. OS Maps shows its summit to be at the OS 1:25000 map 308-metre spot height. The peak is unnamed on the OS 1:25000 and 1:50000 maps, but it seems appropriate to name it after Northmoor Common, which is situated on its east-northeastern flank. Its lower eastern and northern slopes are heavily wooded.

107. Selworthy Beacon, 1,007.87 feet/307.20 metres, SS 918479, an elongated peak, which rises steeply and then fairly steadily from Henners Combe to the north, very gently from a col to the east-southeast and variably from Selworthy and Selworthy Combe to the south and the Bristol Channel and Porlock Bay (via Bossington Hill) to the west-northwest. OS Maps shows its summit to be by the OS 1:25000 map trig point, which is marked as standing at 308 metres. A series of Bronze Age burial cairns lie on the eastern side of its greater summit area; the remains of Bury Castle, an Iron Age hillfort and twelfth-century castle, are situated on its south-southwestern flank; the medieval Katherine's Well is located to the southwest of them; Hurlstone Point can be seen at the end of its west-northwestern flank and Selworthy Sand is found at the foot of its north-northwestern one. Its lower southern and southwestern slopes are heavily wooded.

108. Maundown Hill, 1,007.22 feet/307.00 metres, ST 061285, a south-southeast to east-southeast, clockwise-curving, flat-topped ridge. It rises steeply and then fairly steadily from the Tone Valley to the west and variably from a col (via the hamlet of Maundown) to the north, another to the south-southeast and a tributary dale of the Hillfarrance Brook to the east. OS Maps shows its highest point to be immediately west of the small reservoir marked on its summit area on the OS 1:25000 map towards the northern end of the 305-metre contour ring. Wiveliscombe is situated near the foot of its east-southeastern flank; the village of Langley Marsh is located on its east-northeastern one and its lower western and southwestern slopes are covered by Maundown Plantation. The B 3227 runs across its lower southeastern flank, broadly from west-southwest to east-northeast.

109. [Cutcombe Hill], 1,004.59 feet/306.20 metres, SS 927392, an elongated, ovalish-topped peak on a north to east-curving spur of Lype Hill, approaching two miles to the southeast. It rises steeply and then gently from the Avill Valley to the north and one of its tributary dales to the east and increasingly gently from another to the west. OS Maps shows its summit to be on the northeastern edge of the OS 1:25000 map 307-metre spot height. However, it doesn't have a relative height of fifty feet and is linked to its parent peak by a high col. Its lower eastern and northern slopes are heavily wooded; one or two disused quarries are situated on its north-northeastern flank and the village of Cutcombe is located at the foot of its southern one. The A 396 bends around its western and northern slopes, broadly from west-southwest to east-northeast.

110. Pen Hill, 999.02 feet/304.50 metres, ST 564487, an elongated, ovalish-topped peak, which rises quite steeply and then very gently from a dry valley (via Holes Ash Spring) to the west-southwest, gently and then fairly steadily from another one to the north and variably from a col to the east and Biddle Combe (via Prior's Hill) to the south. OS Maps shows its summit to be at the OS 1:25000 map trig point, which is marked as standing at 305 metres. However, it doesn't qualify as a 1,000-foot hill according to OS Maps, which computes its height as being just 0.88 feet short of the target. The trig point surmounts a prehistoric cairn; the Mendip UHF Transmitting Station is situated at the northern end of its summit area; a round barrow and a Neolithic long barrow are located at its southwestern end and there are a number of other prehistoric remains on its slopes. There is a fair amount of woodland on its southern and western flanks and The A 39 runs across its southern and eastern ones, broadly from south-southwest to north-northeast.

111. North Hill, 998.36 feet/304.30 metres, ST 539514, an elongated, curved peak, which rises quite gently and then very gently from a pool and stream in Priddy Mineries Nature Reserve to the east, gently and then quite gently from a dry valley to the south and variably from an unnamed dale to the west and a col to the north. OS Maps shows its summit to be at the OS 1:25000 map 307-metre spot height, on the top of a Bronze Age bowl barrow. Nevertheless, it doesn't qualify as a 1,000-foot hill according to OS Maps, which computes its height as being just 1.64 feet short of the target. The Bronze Age Priddy Nine Barrows, which run across its greater summit area, and the Priddy Circles, four large and probably Neolithic earthworks situated on its northern flank, are amongst numerous prehistoric remains lying on the hill. Eastwater Cavern is situated on its southern flank; Fair Lady Well and St Cuthbert's Swallet are located on its south-southeastern one; Swildon's Hole Cavern is sited on its western one; the village of Priddy lies at the foot of its west-southwestern one and North Hill Swallet is found on its south-southwestern one. The B 3135 runs across its northern slopes, broadly from east to west.

112. [Weacombe Hill], 997.38 feet/304.00 metres, ST 127400, a small, ovalish-shaped peak on a west-northwest-projecting spur of Thorncombe Hill, to the south-southeast. It rises steeply and then steadily from the head of Bicknoller Combe to the southwest, steeply and then quite gently from a tributary dale of Sheppard's Combe to the east, very gently from a col to the north-northeast and variably from the Doniford Valley to the west. OS Maps shows its summit to be approximately 60 yards southwest of the OS 1:25000 map 305-metre spot height. However, it doesn't have a relative height of fifty feet and is linked to its parent peak by a very high col. Also, despite being marked on the OS 1:25000 map as reaching 305 metres, it doesn't qualify as a 1,000-foot hill according to OS Maps, which computes its height as being just 2.62 feet short of the target. There are at least two prehistoric round barrows situated on its summit area and the A 358 and a railway line run across its lower western slopes, broadly from east-southeast to west-northwest.